Dealing with GREAT AMMO SHORTAGE

AF675655

R.K. Campbell

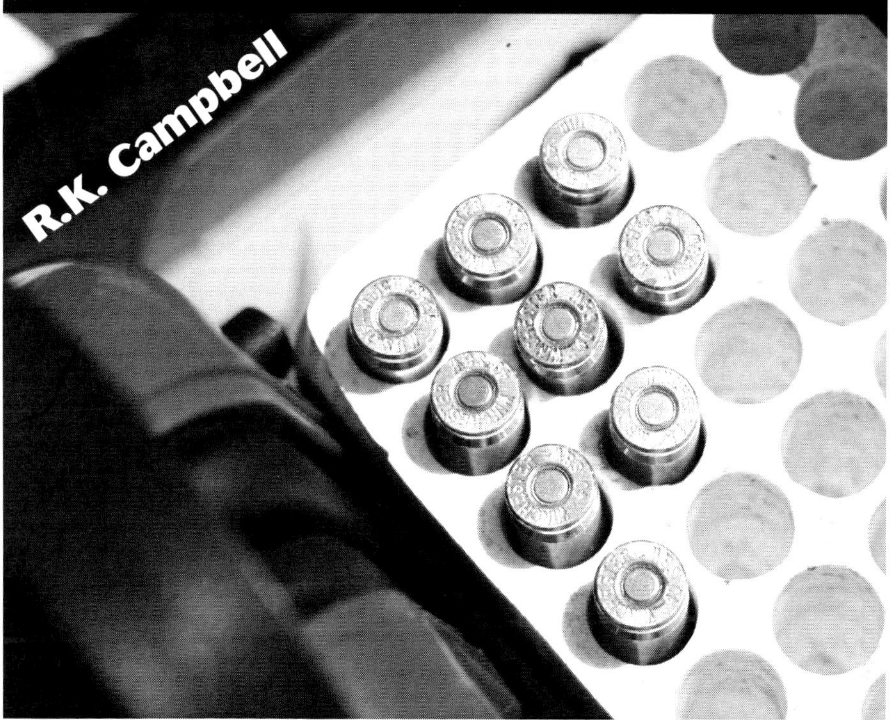

Paladin Press · Boulder, Colorado

Also by R.K. Campbell:
Holsters for Combat and Concealed Carry

Dealing with the Great Ammo Shortage
by R.K. Campbell

Copyright © 2013 by R.K. Campbell

ISBN 13: 978-1-61004-868-2
Printed in the United States of America

Published by Paladin Press, a division of
Paladin Enterprises, Inc.
Gunbarrel Tech Center
7077 Winchester Circle
Boulder, Colorado 80301 USA, +1.303.443.7250

Direct inquiries and/or orders to the above address.

PALADIN, PALADIN PRESS, and the "horse head" design are trademarks belonging to Paladin Enterprises and registered in United States Patent and Trademark Office.

All rights reserved. Except for use in a review, no portion of this book may be reproduced, stored in or introduced into a retrieval system, or transmitted in any form without the express written permission of the publisher. The scanning, uploading, and distribution of this book by the Internet or any other means without the permission of the publisher is illegal and punishable by law. Please respect the author's rights and do not participate in any form of electronic piracy of copyrighted material.

Front cover photos: (alphabet letters) Vrender/iStock; (box of ammo) JoeDphoto/iStock.

Neither the author nor the publisher assumes any responsibility for the use or misuse of information contained in this book.

Visit our website at www.paladin-press.com

• • •

TABLE OF CONTENTS

The author and his wife, Joyce, in Paris in 2012.

I am blessed beyond measure to live completely for and with
what I love best on earth. That is Joyce and my family.

If you love shooting, do more of it.
Work at the shortage smartly, and in times of feast prepare for famine.

··· CHAPTER 1 ···

WHAT IS THE SHORTAGE ABOUT?

The first hurdle in addressing the ammunition shortage we have come to know and hate is to convince the shooter that the shortage is real. It is as real as any economic reality. While gold and silver prices and hog futures may be manipulated, the same is not true of the ammunition situation. It is a true shortage, although some play it to their advantage. There is simply more demand than ammunition.

First, there is a tremendous demand from the military for training, practice, and operational ammunition. Our soldiers are the best trained and equipped in the world, and it takes a lot of ammunition to train these men and women to such a high standard. We also supply ammunition to other militaries.

Second, the requirement for law enforcement personnel has increased dramatically with the increased militarization of the police and the resultant demand for training ammunition.

With the police demanding thousands of rounds per officer annually to train SWAT, SERT, and other specialized teams, and with increasingly more departments demanding handgun qualifications, there is a great demand for ammunition in these circles. As an example, one government agency order recently topped 174,000 rounds of .357 SIG ammunition. That is a lot of ammo, but for training and deployment for thousands of federal officers, perhaps not so surprising after a bit of thought.

This particular order was for the offices of Inspector General and Investigation. While the exact figure is difficult to pin down, it is a reasonable estimate that all federal law enforcement agencies combined demand about 4 1/2 million rounds of ammunition a year. These centerfire rounds are far more complex than the approximately 4 billion rounds of .22-caliber rimfire ammunition produced each year. There is a lot of testing and development involved in working up a new loading and then retesting and developing each lot of ammunition. This requires a great deal of raw material, including brass, copper, lead, and in some cases other materials. No wonder the makers are looking to such alternatives as aluminum and steel for cases and various alloys for bullets.

It stands to reason that the .45 ACP cartridge, left, demands more raw material, more expense, and more storage space than the lowly .22 LR, right.

Another consideration is the great diversity in ammunition. There are often a half dozen loadings in the popular calibers from a single ammunition producer. As an example, in .45 ACP caliber, Winchester offers the standard 230-grain ball load, a 185-grain Silvertip, and the 230-grain PDX, as well as the Ranger police line. Cor-Bon is a smaller company—almost a custom line—but offers a surprising array of loadings in .45

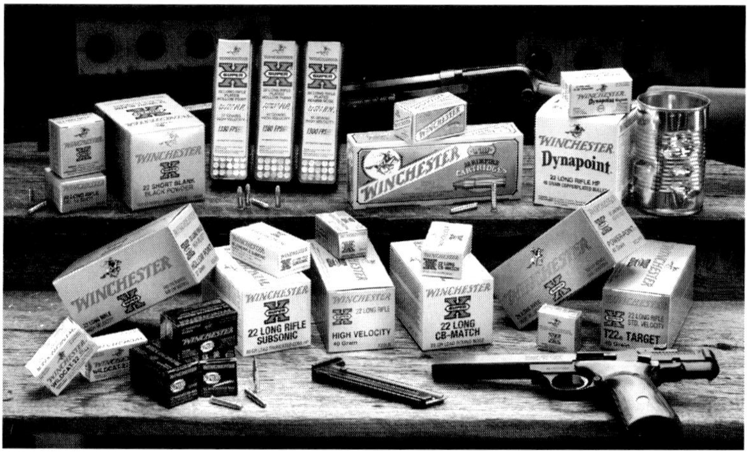

The .22-caliber rimfire cartridge is available in a bewildering array of loadings. Whatever other loads you may choose, the .22 should be top on the list for practice and training.

ACP, including a 160-grain DPX "short-barrel load," the 185-grain DPX, 165-grain Pow'RBall, 165-grain JHP, 185-grain JHP, 200-grain JHP, 230-grain JHP, and Glaser loads . . . not to mention the target-grade Performance Math loads. This is a lot of diversity, yet the company maintains its reputation for custom-grade quality and performance.

Hornady is one of our most respected makers. The company has a tremendous lab for developing projectiles, and its techs get it right. When Hornady introduces something like the Critical Defense bullet, other makers must catch up. And this catch-up may mean months of expensive research and development. Since these companies do not have the machines running 24 hours to produce certain loads, catch-up isn't always feasible. Often the machines turn out millions of rounds in batches, and then the heads are changed to another caliber or the machine is set up for the same caliber but a different loading. It only makes sense that the machines run wide open in turning out millions of rounds in calibers that are in great demand, such as the 9mm Luger and the .45 ACP.

Less-popular cartridges, such as the .32 Smith & Wesson

Long, are going to be a low priority. In some cases these little cartridges with their round-nose lead bullets are twice as expensive as modern jacketed rounds for the 9mm even though they use far less material. It doesn't make horse sense until you figure supply and demand, with the more popular rounds selling at 100 to 1 or maybe 1,000 to 1 compared to the .32. If you own an oddball caliber, such as the .32 H&R Magnum, the .44-40 Winchester, the .45 Long Colt, or the .455 Webley, you really need to stock up on ammunition. Some of these calibers can't be obtained in a casual trip to the gun shop.

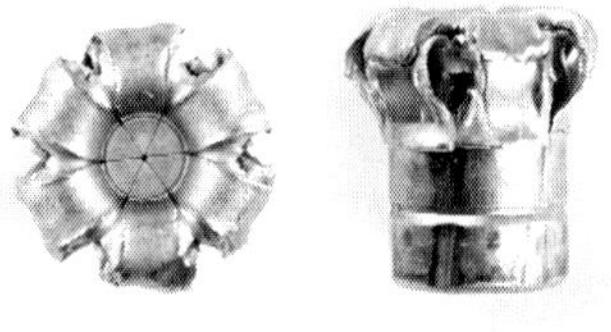

The Hornady Critical Defense (CD) bullet is among the most effective ammo developments of recent years.

Scrounging rounds is one option, but it is even better to learn how to load your own. Otherwise you will be caught short in an ammunition shortage. During the shortage some calibers are going to be hard to come by, while others will be manageable with some research. Sometimes what seems an oddball may be available with a bit of thought and loading tools. As an example, a cartridge such as the .45 Auto Rim makes sense because while the cartridge case is different, the components are the same as the common .45 ACP. Even .45 ACP dies may be used to load the cartridge—all you need is a .45 AR shell holder. The .38-40 and the .32-20 are far different propositions. They are going to be more difficult all the way around. A worthwhile book to have is *Cartridge Conversions* by George Nonte, which tells how to take existing cartridges you have and modify the rounds (casings) to fit a different-caliber weapon.

Another factor in the ammunition shortage is a genuine shortfall in raw materials. At times copper and brass in the correct alloy and in sufficient quantity have been difficult to obtain. It appears that lead is not as scarce as the other raw materials, but, just the same, lead is becoming more expensive than ever. Manpower and machinery is never inexpensive. As a result of these various considerations, ammunition prices

Just take a look at some of the available 9mm Luger loadings. All these are good choices for one purpose or another, but you have to decide on your choice and plan ahead.

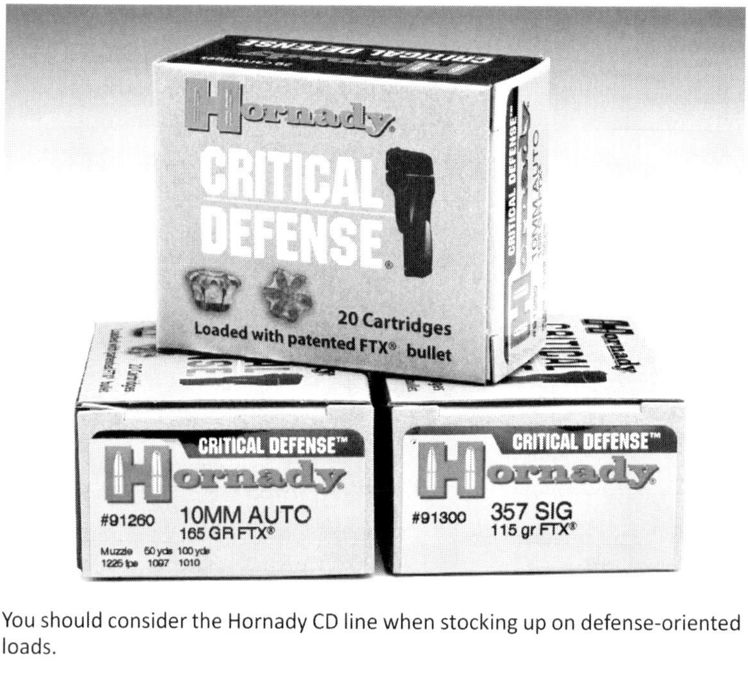

You should consider the Hornady CD line when stocking up on defense-oriented loads.

The .44-40 is far from the most widely distributed cartridge, but Black Hills loads the round. The .44-40 is easy to shoot well, is accurate, and can be very effective.

rose during the shortage of 2009 with some prices rising 40 percent. There was a great deal of price gouging, and all I can say is to remember the shops that were fair and the ones that were not. The makers did not price gouge; the ammunition went up just a little. It was simply hard to find due to heightened demand. Shops charging a 100-percent markup were the price gougers. While they made an unscrupulous short-term profit, in the long run their unfair practices should be remembered. Most of the makers held their prices, but some were forced to raise their price due to the increasing cost of raw materials.

A not inconsiderable ingredient in the mix, and certainly a catalyst for the inflammatory rise in prices in 2009, was the election of Barack Obama and his appointment of rabid anti-gunners. Despite the fact that they haven't done much to further their agenda, at least as of this writing, their history speaks for itself. Their thoughts and deeds in the past were frightening and still are to anyone who loves liberty. Some of these fears are well founded, and it remains to be seen what mark his ultraliberal regime will leave on America.

Some fears are irrational to an extent, but a perceived and a real concern produce the same end result in this case: panic buying and hoarding of ammo. Folks who previously purchased perhaps one or two boxes of ammunition a year were suddenly desirous of 10 boxes or a case (1,000 rounds) just in case the ammunition could not be obtained elsewhere. Free men and women made certain they had the guns and ammunition they desired. The result was a pandemic. Folks were buying guns and 10 to 20 boxes of ammunition. Some were not real gun folks, like you and I, and after due consideration and when the credit card bills came due, these individuals often unloaded these items at the same shop at which they had been purchased. They sometimes took a beating, and astute traders benefited from the results.

And the citizen is not alone in suffering from extra expense. The United States, in an embarrassing situation, paid an estimated $10 million tariff over the cost of self-producing ammunition to obtain good quality ammunition from Israeli Military Industries. At least the money went to a friend and ally, but the fact that we had to go this route was embarrassing. Just the same, the government was not alone.

The shortage has overwhelmed many of us. In the end the ammunition shortage isn't a myth. Simply, it may be summed up by stating that the demand for ammunition has outstripped manufacturing capability. While the manufacturing ability of American companies has been greatly expanded in most cases, few have obtained extra machines or built larger plants. Rather, they have put extra shifts to work to try to keep up with increased demand. Police agencies are now ordering 120 days ahead and placing larger orders on a more regular basis. You should do the same in order to keep your ammunition supply current, but on an individual basis there are smart cost-saving cuts to consider.

Another factor that is driving up ammunition prices is that people are simply shooting more. Sure, some are hoarding, but

The days of taking a good rifle to the range and working it out with a variety of loads may be over. Better do your testing at the first opportunity and cache the load that best suits your rifle.

When shortages exist, brand loyalty is difficult to maintain. All these loads have given excellent service and good accuracy.

The rather pedestrian .30-30 Winchester may take a new lease on life with a loading such as the Winchester Power Point. This is an effective caliber up to 100 yards, making it well suited to backwoods use. It is also inexpensive.

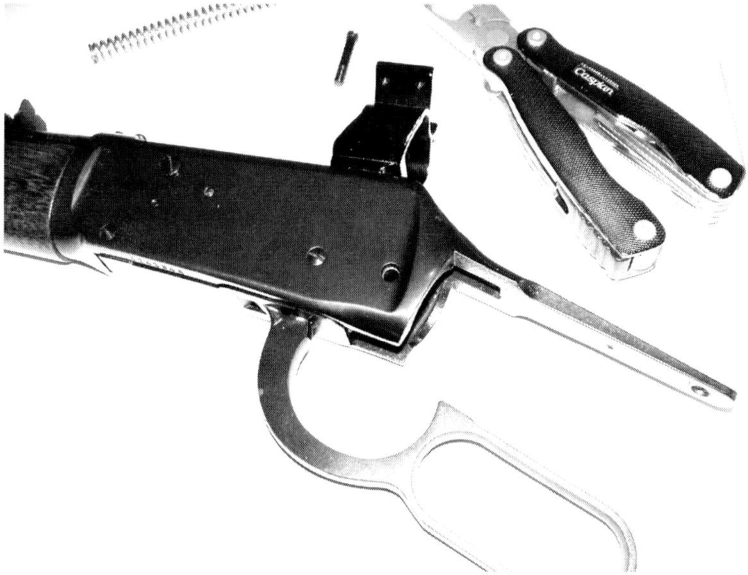

Be certain to choose a rifle you are able to maintain in the long term. The Winchester Centerfire .30-30 is simple, reliable, and powerful.

others are shooting a lot of ammunition. The number of men and women engaged in competition shooting, IDPA matches, training for concealed carry classes, and recreational shooting is at an all-time high. The ammunition companies simply did not see this coming, at least that's what many of us have concluded.

Another point that is arguable by some but which I feel qualified to comment on is ammunition quality. All ammunition is not created equal. If the ammunition is less expensive, some corners have been cut. When the choices are few, we often are forced to buy what we would find, and it is unfortunate that sometimes brand loyalty is forced to the back burner during the shortages. You just can't always find the preferred brand. Many shooters favor one maker over the other for various subjective reasons. If you like the Winchester Power Point, you are in good company, but just the same there are merits for the Federal Fusion line as well. The problem is you cannot always obtain the quantity of ammunition you need if you adhere to brand loyalty.

These lead bullets are high quality, consistent, and accurate.

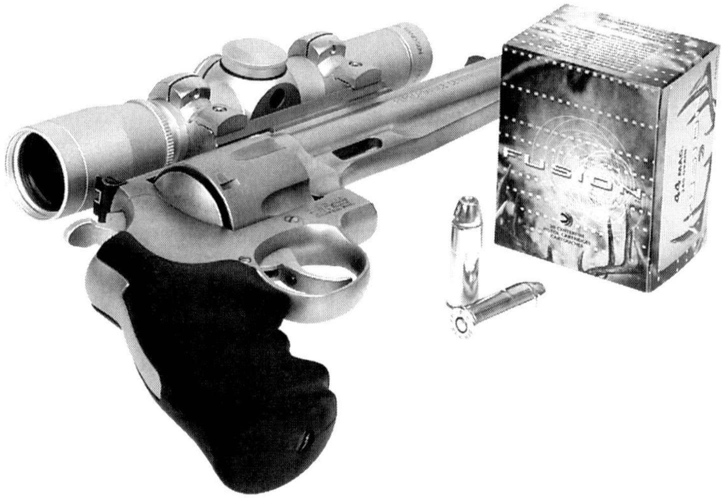

If hunting is your game, stock up on the chosen load. Such low-production items as Fusion hunting loads will be the first to disappear during a shortage.

The difference in "red box" Black Hills loads is detectable in only the most accurate rifles. For general use, the blue box remanufactured loads are just fine.

The Black Hills remanufactured line, using fired brass that has been cleaned and reloaded, is most often indistinguishable from the new line in accuracy and reliability.

Sometimes you have to make concessions and grab what you can find. As an example, like most of you I use quality hollowpoint ammunition in my carry guns. I have proofed these loads in my personal firearms. But there is no real reason to use these loads for everyday practice. Full-metal-jacketed (FMJ) and lead bullets are perfectly acceptable for practice. Simply be certain that the inexpensive practice loads have the same point of aim and point of impact relationship and you will have a meaningful practice load.

As the ammunition situation has tightened, quite a few manufacturers have risen to meet the demand, and some offer good ammunition at a fair price. As an example, Black Hills Ammunition, a respected maker of excellent service-grade ammunition, offers first-quality remanufactured ammunition. I have used thousands of rounds of this ammunition for training purposes. Another maker, Buffalo Bore, offers specialty loads for hunting and personal defense, including a Tactical line. Both companies offer good products.

Other companies seem to have cut too many corners in quality control. The loads may be unreliable in ignition, cycle reliability may be questionable, and in some cases the loads are actually dangerous to use and fire. Read on and I will address simple solutions to a complex problem.

· · · CHAPTER 2 · · ·

A CARTRIDGE AND ITS COMPONENTS

To learn how to store, fabricate, and handle ammunition, you need to understand cartridges and their construction. The cartridge case is the main element, holding the other vital parts of the cartridge, including the powder charge. There is a hole in the base of the cartridge case that accepts a flush-fitted primer. The primer acts as the igniter or spark plug to set off the powder charge. Pressed or crimped into the case mouth is the projectile. The projectile may be properly called a bullet. The cartridge is sometimes called a round. The cartridge is often incorrectly called the bullet. The bullet is the projectile not the entire cartridge.

The cartridge case is manufactured from brass in most cases, but there are alternate materials in use as well, including steel and aluminum. The purpose of the cartridge is to seal the pressure in the cartridge case and direct the gas formed by powder ignition into the bore of the firearm. The primer contains a small amount of compound that is ignited by the force of the firing pin or striker impacting the metal priming cup. The priming cup is driven against an anvil, and the primer compound is ignited. The flash or fire travels through the flash hole in the bottom of the case and ignites the powder. In a rimfire cartridge, the priming compound is contained in the case rim, and it is placed all the way around the case rim. The priming compound is ignited when the priming compound is crushed against the brass by the firing pin.

There is one flash hole with the standard primer in use in most American cartridges, but the Berdan primer is also used worldwide. The Berdan type fea-

Above: A handgun cartridge is more complex than some realize. This Hornady cartridge is built with something called "cartridge integrity."

Left: A .22-caliber cartridge is made of a drawn brass cartridge case with fulminate compound in the base for ignition. The soft lead bullet is a heel type that is pressed into the cartridge neck.

tures two flash holes in the case. While an argument may be made for the Berdan primer, it is not easily replaced and Berdan primed cases are generally regarded as nonreloadable.

The primer ignites the powder after the primer is struck by the firing pin. As the powder ignites, the pressure from the gas then seals the cartridge case against the chamber walls in the firearm. These burning gases press the projectile forward in the barrel. It is not an explosion but rather a controlled burn of gas. A detonation would wreck the firearm. As pressure builds, the cartridge case expands temporarily to press and seal against the chamber walls. Once the bullet exits the muzzle, then pressure dissipates and the cartridge case no longer adheres to the chamber wall.

As you may surmise, the cartridge case must be strong but also maintain a degree of elasticity. Brass is thus far the most successful material for use in cartridge cases. Brass is easily worked, resists pressure well, and is not brittle. A brass cartridge case may easily be reloaded. Steel is used in some applications because it is less expensive and more readily available. Steel cases may sometimes be harder on the extractor, so care must be taken in their use. In fact, there is some debate about the use of steel cases. Some base their opinion purely on the fact that steel cases are inexpensive; therefore, they must be inferior, they reason. This isn't necessarily the case, but they are certainly different.

You have to look at the facts. We are all entitled to an opinion but not our own facts. It seems that the firearms that are originally designed for steel cases fare better as far as extractor wear with steel cases. These firearms include the AK-47 variants. Others designed for brass-cased cartridges may not have an extractor of sufficient hardness to deal with steel cases in the long run. Steel cases are about 20 to 25 percent harder than brass cases. Steel cartridge cases are far more resistant to wear but, on the other hand, are usually regarded as nonreloadable. This may seem contraindicated, but sizing dies would have to be redesigned to take steel cases. I can't think of any gun where the extractor is not an inexpensive and easily replaced part.

Therefore, if you shoot steel cases, lay in some spare extractors.

And then there is the problem of the Berdan-primed cartridge case. To remove a Berdan primer, a standard decapping pin would not be applicable. There are two flash holes that offset the Berdan primer, and the anvil is actually part of the cartridge case head. The only way to remove the primer is by water pressure. Fill the case with water and drive something such as a dowel that is a tight flush fit into the case and you may be able to pop out the Berdan primer. Back around the turn of the 20th century, there was a tool that hooked in the primer, punctured it, and then removed it, but now they're antiques and I don't know of any currently on the market. I have never actually seen one. As for the hydraulic method, I have done so experimentally, but remember that CCI purposely designed its Blazer aluminum case line with a Berdan priming system in order to ensure that folks would not be trying to reload aluminum cases. The problem with aluminum is that it has poor memory and is less malleable (that's why you see brass springs but not aluminum unless an exotic alloy is used).

This shows pistol cartridge cases before and after firing. The cartridges at top are standard boxer-primed cases; the cases at the bottom are Berdan primed.

The main point to consider when stockpiling ammunition is that there are advantages and disadvantages to steel cases. The fact that they are pretty rugged is an advantage. (We will discuss the advantages of Berdan primers in the chapter on primers.) However there are also disadvantages to steel cases including the fact that this hardness may result in greater wear on the magazines, feed ramp, chamber, and particularly the cartridge-case extractor. It is very difficult to gauge the extent of the wear increase, and chances are it will not be a factor before the

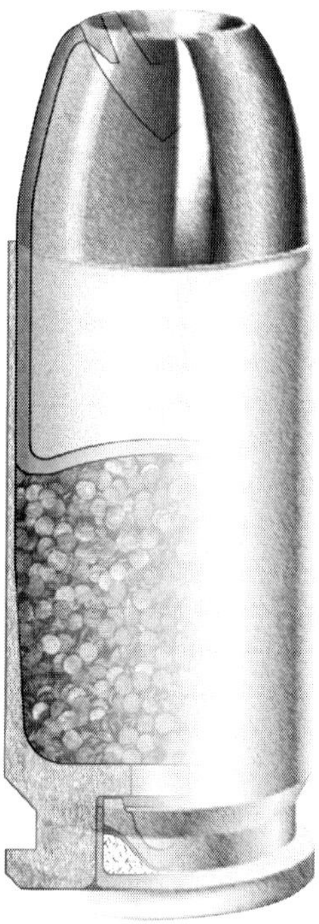

This is a CCI Blazer cartridge. Note the powder, jacketed bullet, aluminum case, and the Berdan priming method.

10,000-round mark. But it is a consideration. If you are purchasing a few thousand rounds for the great what-if, then steel-cased loads are certainly an option. Remember, however, if you stockpile steel-cased ammunition, rust is a concern. The lacquer-coated rounds are not coated for feed reliability but rather to decrease the risk of corrosion. A number of American companies now offer quality steel-cased ammunition. A forerunner in the domestic use of steel cases is Black Hills Ammunition of Rapid City, South Dakota.

Another material that has seen some use is aluminum. Speer CCI offers the Blazer line with aluminum cases for economy. The famous A-10 Warthog tank buster with its giant 30mm gun uses aluminum-cased cannon shells, so aluminum-cased rounds are military worthy. Military sources tell me that with conventional brass-cased shells, the A-10 would be too heavy to fly.

Whatever the cartridge case material, the cartridge must be produced with dimensional integrity in order to properly feed, chamber, fire, and eject.

The powder charge is contained in the cartridge case. The powder charge is measured in grains, with 7,000 grains to a pound, and can range from just over a grain in weight in the .25 ACP to well over 50 grains of powder in a .300 Magnum. The powder may be very fast burning for use in small-capacity handgun cartridge cases, or slower burning for Magnum revolvers and extra slow burning for use in rifles. The bullet weight is also measured in grains. Bullets range from the 17-grain pill of a .17 rimfire to well over 500 grains in a .45-70 rifle cartridge.

Bullet diameter is measured in inches in hundredth-of-an-inch increments, such as .223 Remington or .45 Colt. The metric system is also used to measure bullet diameter, such as the 9mm Luger and the American 10mm automatic. The metric system is more specific in measuring cartridge cases as well. As an example, the 9x18 Makarov has a shorter case than the 9x19 Luger, as the 19 in the measurement designates the length of the cartridge case.

It is important to realize the differences in the types of cartridge cases. Most rifle cartridges are bottleneck designs for good feed reliability and chambering. These include the .308 Winchester and .30-06 Springfield. The few straight-walled rifle cartridges, such as the .30 carbine, are less-powerful examples. Or there are the larger, older rounds such as the .45-70. Feed reliability wasn't important in single-shot rifles. There are a number of bottleneck pistol cartridges, including the 7.62mm Tokarev and .357 SIG. The 9mm has a slight taper, while the .44-40 Winchester, originally designed for use in lever-action rifles and adapted to handguns, has a more pronounced taper short of a true bottleneck. Most pistol cartridges are straight walled.

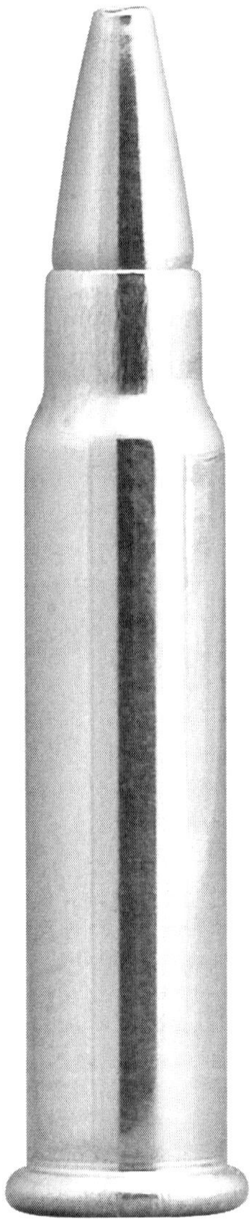

The .223 Remington is among the most popular of all calibers in America. It is accurate, powerful, and compact.

While there has been some success in adopting cartridges to both pistols and revolvers, the types remain distinct. The revolver cartridge chambers on a step in the cylinder. The cartridge case

Captain Matthew Campbell has two rounds in the air with the Ruger .223. It takes a lot of practice to get it right.

rim accomplishes headspace. The automatic pistol uses a cartridge that headspaces on the cartridge mouth. (There are a few exceptions, such as the early .38 ACP pistol. The .380 ACP cartridge—like the .32 ACP also designed by John Browning—was an early attempt to have it both ways: these are *semi-rimmed* rounds that have a narrow rim to help accomplish headspace, but not enough to hamper feeding, plus a groove for the extractor to fit into and grip.)

In cartridges designed for automatic loading without a rim for the extractor to grip, the case head features an extractor groove that allows the extractor to grab the case and remove it from the chamber. There are many calibers, and the longevity of the cartridge and the shooter alike demands that the proper cartridge case be used in each firearm. If you are not a handloader, then you should not be overly concerned with purchasing brass-cased ammunition. You may not be using the cartridge case to reload, so economy is the goal. If you are a reloader, then by all means obtain brass-cased new cartridges that, once fired, may be used for reloading. Steel cases cannot be reloaded easily. Also, they do not give as much as brass on firing, and there is some blowby of gas back into the chamber, resulting in a dirty chamber, especially in rifles with fluted chambers, such as the World War II Russian Tokarev rifle, the modern HK G3-type rifles, or the Canadian Ross rifles that had chambers enlarged to facilitate extraction. Also, steel cases will rust almost immediately. They are usually coated on the outside to prevent corrosion, but the interior will rust quickly once the cartridge is fired.

Fired cartridge cases are not the only source of cases for reloading. Some of the major makers offer cartridge cases for the reloader. These cases are often in short supply, and frankly they are so expensive you are about as well off to simply purchase economical ammunition, such as Winchester USA, and fire it for the cartridge cases. I have actually seen Winchester USA in .308 caliber available as loaded ammunition for less than a box of 20 new empty cartridge cases from a competing maker!

For the handloader wishing to get started the right way, Starline Brass offers first-quality cartridge cases at a fair price. I have used thousands with good results. Many ammunition makers use Starline Brass. Starline has changed the industry. Many years ago Super Vel Cartridge, a leader in both quality and performance, was forced out of business simply because it could not obtain enough quality brass cartridge cases to fill its many orders. Super Vel was at the mercy of a couple of makers of brass cases that were also their competitors in the ammunition field. Today, with the independent maker Starline supplying millions of bass cases, the situation for firms such as Cor-Bon is far brighter.

When it comes to quality, all cartridge

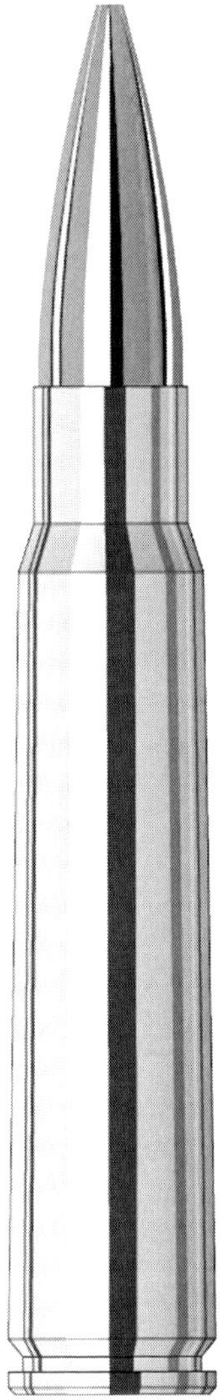

A centerfire rifle cartridge is fairly complex. The projectile demands high manufacturing quality because the cartridge operates at high pressure.

brass is not the same. I have examined brass cases that could not be properly resized, with flash holes off center or indifferent, and some that simply would not properly chamber. Be careful what you use.

I will discuss powder and bullet selection at length in the following chapters. Remember, before the invention of the modern boxer primed cartridge, ignition success was unpredictable. A flint, a percussion cap, or a spark was needed to ignite black power. The percussion cap was a great improvement over the flintlock, and the rimfire cartridge an even better one, but the modern centerfire cartridge was essential to repeating firearms.

Paper cartridges were the first step in the road to combining the charge and cap in one package. The rimfire cartridge and the centerfire cartridge were tremendous advances in firearms operation and reliability.

• • • CHAPTER 3 • • •

STARTING LOCAL

This chapter was written by Robert Alan Campbell.

The search for ammunition begins with the purchase of the first handgun or rifle. Depending on where the gun was purchased, normally most buyers would tack on the sale of the first box of ammunition for the caliber. But once that first box and the initial excitement of the new purchase ebb, the search begins in earnest. You either have the bug for shooting and learning to shoot well, to shoot straight, and to enjoy the sport, or you don't. Since you are reading this book, I assume you have the bug!

The best places to start are right in your own backyard. Walmart carries thousands of rounds of ammunition in brands such as Winchester, Tula Ammo, Remington, and Federal, and my regional stores recently started carrying personal-defense loads. Many a time while gathering the kids' new school clothes or house supplies, I've found myself down the sporting goods

Practice leads to proficiency. When ammunition is a nonrenewable resource, you are less likely to pull the trigger as often on the range.

Decide what you are doing with the ammunition before you purchase any. For a combination of personal defense and deer-sized game, the Black Hills 62-grain load simply cannot be bested.

aisle peering into the glass case, checking out each caliber, grain, and bullet type, and comparing prices. The handgun calibers currently stocked include .22, .380 Auto, 9mm, .38 Special, .40 Caliber, .357 Magnum, .45 Auto, and even .44 Magnum. But it doesn't stop there. Included on the shelves are rifle calibers in .223, 7.62x39, .308, .30-30, .30-06, .410, and 20- and 12-gauge shotshells. Of course, caliber availability might vary for each store or region. Walmart has helped me on more than a few range days when in my haste to pack and get to the range I'd forgotten to grab .22s or even .38 Specials! At those moments, it wasn't difficult to find a brick of Winchester Super-X in .22 Long or even Federal 230 grain .45s on a whim. Although Walmart stores don't usually have three to four aisles of ammunition, they do have a great selection of reputable ammunition in most common handgun and rifle calibers and shotshells.

One note about Walmart: in January 2013, the company announced that it would begin limiting sales of ammo to three boxes per customer per day because of the shortage. Factor this into your resupply plans.

This bulk offering from Winchester is two boxes in one—a good deal when found on sale at Walmart.

If light recoil, light weight, and finesse mean anything, the Walther PK380 is a great pistol. Traditionally, the .380 ACP is among the hardest rounds to find during a shortage, so stock up now.

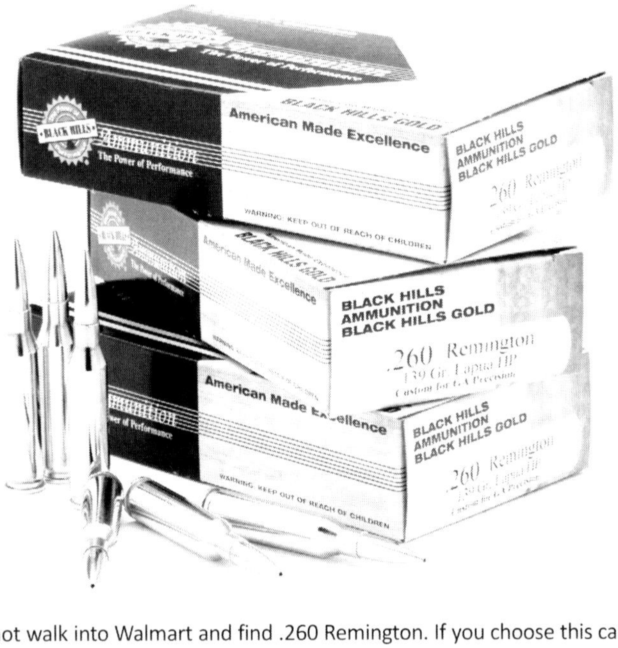

You cannot walk into Walmart and find .260 Remington. If you choose this caliber and Black Hill's top-rated loading, then you should stock up on it as soon as possible.

Glock, Galco holster, and Cor-Bon ammunition. But what if the ammunition you need isn't on the shelves? Will you take what you can find?

What about other brands that are not stocked at Walmart? This shifts the search toward pawnshops and gun stores. Almost every pawn store that has handguns and rifles for sale tries to carry a greater amount of unique ammo and brands and calibers. After all, if the store has a Colt Series 80 in .38 Super available for purchase, it's in their best interest to carry at least a few boxes of .38 Super to accompany the handgun. Fiocchi's 50-round boxes are of high quality in the European standard and usually affordable.

Most people purchasing their first handgun normally buy one box of range ammo and one box of personal-defense ammo. For this reason, it would be hard not to find exactly what you are looking for in a pawn or gun store. They tend to keep a fair selection for any caliber stocked and ready for the next customer. Here you could find brands such as Fiocchi with the X-Trema line, Wolf, and especially Hornady and Speer. One of my favorite boxes of ammo to search for locally is the CCI Shotshells in .45 Auto and other calibers. Although they can be purchased on-line from a retail supplier, it's much cheaper to walk into the store and purchase these without paying shipping and hazmat fees for only 10 rounds.

When I asked at the local pawn and gun shops, while purchasing primers, if they had any shell holders or cast bullets, I was gladly redirected to Gander Mountain. Currently, Gander is the nation's largest provider of outdoor and hunting supplies and other accessories, having 113 stores in 23 states, with new stores recently opening in North Carolina, Alabama, and Indiana. The chain carries a vast number of calibers, brands, and varieties from each manufacturer. Most of the time, you will find their aisles stocked and even bulk cases lined with handgun, rifle, and shotshell offerings. For the range jockey in all of us, bulk cases of ammo are a must if you love to shoot! In addition to factory ammo, Gander Mountain also carries reloading supplies and even some equipment. Their stores have one of the largest selections of primers I've seen in a retail store. From small pistol to large rifle, they have earned a spot on my list of local suppliers. Their smokeless powder selection is not limited either. Filling in the cabinets are common brands like Hodgdon, IMR, Winchester, and Alliant. If they don't have a certain type, they will be glad to add it to their next order. For most of us with a variety of calibers, a certain variety of powders is a must. Also on the reloading aisle are bags of unfired, unprimed brass and even Hornady's XTPs in most caliber sizes.

Some areas don't have the advantage of a mega outdoors store for ammunition and reloading items, but it's still not that far-fetched. A quick map search for the region shows the closest city with a similar retailer. Once the price of shipping and delivery times are weighed against the price of gas and drive time, anything is achievable! Some areas with big cities usually have annual or even quarterly gun shows, which are excellent venues for observing and possibly purchasing unique arms. Often vendors will bring in loads of factory ammo they no longer require and offer it at a discounted price.

I would stress to only purchase ammo that you are absolutely sure is factory ammo. Older-type military surplus ammo or even reloads could damage your firearm if not properly inspected and possibly injury or death could occur from overcharged or improperly sized reloads. If you're unsure, it's not worth the purchase to tempt fate.

On some occasions, reloading supplies—such as new containers of powder, once-fired brass, opened bulk boxes of various bullets—can be found. Even die sets like RCBS and Lee show up every so often. Most of the time these items are used but in good condition. The previous owner usually has no further use for the die set after they sell off the firearm in that caliber, which seems to make such equipment finds a steal. But as with any purchase, be sure to ask questions and inspect the equipment before purchasing it to ensure reliability.

For the reloading savants among us, purchasing reloading supplies is an ongoing search. Once you find suppliers locally, it is much more convenient. Luckily in my area, there are three separate pawn and gun stores that carry a few different supplies for reloading. All three of these stores stock a supply of small pistol primers, large pistol primers, a few small magnum primers, once-fired brass, and a handful of essential standard powders. They all have been added to my checklist of local suppliers. If I happen to be traveling into town, depending on which side I'll end up on, I always check the primer and powder supplies at home before leaving.

Because of hazmat and shipping fees, ordering primers and powders from suppliers by mail is a shock! Local stores under-

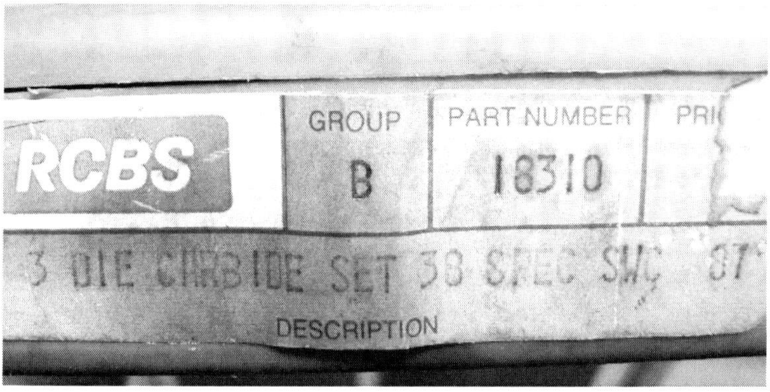

These RCBS dies have been around for more than a decade, and they still work just fine.

stand this concern and stock standard primers and common powders, accounting for the cost of these fees. For a convenient up-charge, these items are available at an economical price for purchase. When comparing the shipping prices and hazmat fees of these reloading items, sometimes you can easily save in excess of $20–30 by buying supplies locally, and you don't have to wait for the supplies to arrive!

• • • CHAPTER 4 • • •

ON-LINE RESOURCES

This chapter was written by Robert Alan Campbell.

Once the local search for ammunition and supplies is either won or exhausted, there are other retailers to consider as well. Most companies these days include a catalog, on-line store, or both as part of their marketing effort. This helps stores to reach a broader base of customers, thus increasing sales. Things to be leery of while making on-line purchases are increased shipping cost, hazmat fees, and even additional handling fees. Still, there are many reputable on-line companies that ensure that they keep their prices and shipping and handling fees reasonable.

Most of the time, such websites require that you create an account with your name, date of birth, and shipping address, but some places will allow you to continue with your purchases as a guest. One of the advantages of on-line catalog ordering is the plethora of choices local stores can't match. It's hard to recall how many of my phone calls and visits to various local

The author looking for best buys on-line from the comfort of home.

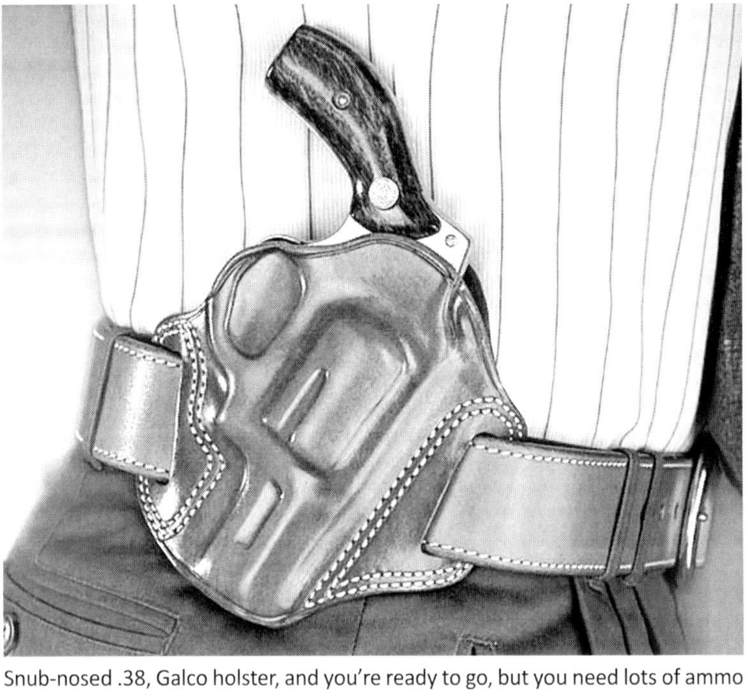

Snub-nosed .38, Galco holster, and you're ready to go, but you need lots of ammo to be proficient with this gear. Buying on-line can be a convenient option.

stores turned up empty when I was looking for a specific bullet weight in caliber or maker. Once you turn your attention to on-line ammunition stores, the search engine goes wild, and the game is on.

Whether you're purchasing manufactured ammunition, reloading supplies, or even shooting accessories, on-line shopping yields many positive products if made through a reputable and trustworthy company. Purchasing items that require hazmat fees, such as primers and powders, increases the cost of those items. Buying bulk amounts of ammo at discounted prices could save hundreds of dollars over time. Most companies carry a wide range of ammunition in multiple calibers to accommodate everyone's specific shooting needs, which makes using on-line store a pleasure.

CHEAPER THAN DIRT

Cheaper Than Dirt was my first on-line experience ordering ammunition. The company has an extremely large selection from multiple manufacturers to meet any type of shooting

This teen is hefting her Remington .223 rifle. She isn't about to practice with the hard-kicking .308, and .223 is cheap and plentiful from Cheaper Than Dirt on-line. At least, as of this writing, it was.

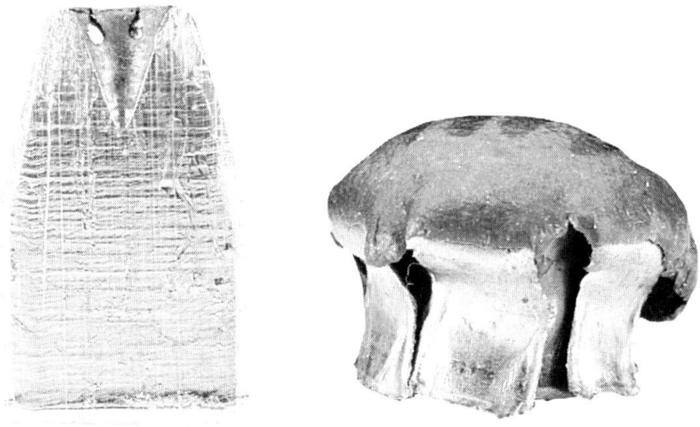

Hornady TAP bullets are among the best available for personal defense, but sometimes they are in short supply. If you accept no substitute, then stock up now.

Captain Matthew Campbell is testing the Hornady TAP .223 and finding it well worth stocking up on!

Winchester is an old-line company that has raised the bar on .223 rifle performance. Stock up now or forever hold an empty piece.

needs (including cartridge adaptors). Normally on the website or in its monthly catalog, you will find weekly and about-to-be-discontinued items. Carrying items from such manufacturers as Hornady, Federal, Sellier & Bellot, Speer, Fiocchi, Winchester, and even Dynamic Research Technologies (DRT), Cheaper Than Dirt offers an abundance of brands for a variety of prices. Bulk cases of surplus .308, .223, 5.56x45, 7.62x54, and .45 Auto are listed in the monthly catalog and on-line homepage, but inventory is not limited to these calibers. Aside from ammunition and almost everything else you can think of, Cheaper Than Dirt also carries new unprimed brass and handgun and rifle bullets for the reloaders. Featuring bullets from Barnes, Speer, Hornady, Lyman, and Winchester, the stock is quickly replenished from multiple warehouses.

GRAF & SONS

Another trusted source is Graf & Sons. If finding primers or powders in your area is difficult, this company has the means to accommodate you. A quick search of primers turns up such brands as Winchester, Federal, Remington, CCI, Sellier & Bellot, and Wolf in large, small, and magnum. Listed alongside the primers are such name-brand powders as Hodgdon, IMR, Alliant, Vihtavuori, Winchester, Ramshot, and Accurate, to name a few. It's easy to find the right powder and primer combination that suits your firearm with so many choices at Graf & Sons.

Because of current hazmat fees from shipping providers, expect to add $27.50 to hazardous materials packages, but the good news is that you can get both products (up to 50 pounds) shipped in the package for one fee. To try to increase sales and productivity, Graf & Sons currently charges only $5.95 for a handling and stocking fee and ships all nonhazardous items over $50 for free within the contiguous United States.

This is a plus since the company offers many bullets in many calibers and quantities for reloading. Some listed manufacturers are Barnes, Speer, Prvi Partizan, Winchester, Remington, and Nosler. Graf & Sons is not shy about showing off its new brass with brands from Starline, Federal, and Armscor, and it even occasionally sells used brass.

Don't fret if you haven't gotten around to learning to handload your own just yet; Graf & Sons also has you covered. With weekly specials on rifle and handgun ammunition ranging from large-quantity-discounted items to bulk buys, this is a great place to search for your ammunition needs whether for plinking, target shooting, or personal defense.

MIDWAYUSA

MidwayUSA is another company that offers one-stop shopping for ammunition and firearms. It carries a wide selection of ammunition, magazines, holsters, targets, and other shooting accessories. Midway has an impressive smokeless powder selection with such brands as Hodgdon, Ramshot, Vihtavuori, Blackhorn, and American Pioneer, to name a few. Another find from searching its catalog is its stock of bullets available from Oregon Trail Bullet Company. Oregon Trail's Laser-Cast bullets are fused with silver alloy and engineered to be a match precision bullet at regular cast-bullet prices, satisfaction guaranteed. My experience with the Laser-Cast line has been nothing less than extraordinary. Their superior hardness allows the bullet to be pushed to a higher velocity without concern for leading. They are an absolute must for any reloader. Most of the time it simply isn't worth getting set up to cast your own bullets, not even counting the inevitable learning curve before you begin to turn out acceptable bullets.

MIDSOUTH SHOOTERS SUPPLY

Midsouth Shooters Supply carries a large amount of remanufactured ammunition. Brands vary from Ultramax Remanufactured Ammo, Blazer, Hornady, Speer, and Buffalo Bore Ammo. Be sure to check out its clearance selection for great deals and discounts. Along with loaded ammo, Midsouth also offers reloading supplies and equipment. Hodgdon, Vihtavuori, IMR, Accurate, and Alliant powders are stocked in 1-, 4-, and 8-pound containers. Accompanying the smokeless powders are CCI, Federal, and Remington primers.

MAGNUS BULLET COMPANY

Record-setting bullets from Magnus Bullet Company are essential when stocking up for target, competition, cowboy, or fun shooting. Magnus currently produces handgun and rifle calibers in more than 50 styles in hard-cast, swaged-lead, and jacketed bullets. Magnus also produces some of the best match-grade hard-cast bullets I've ever used. In particular, the 100-grain pill in .32 caliber will maximize the .32 Magnum like few others.

CHAPTER 5

HANDLOADING

Too many shooters think of handloaded ammunition as a second-rate choice, something less effective than that produced in a factory. Nothing could be further from the truth. If you 've considered handloading but never attempted it, you need to stop thinking about it and start doing it!

The process isn't that complicated. It begins with the spent cartridge case. With a sizing die mounted on a press, you resize the case to its original dimensions since it expands a bit on firing. As the case is resized, the depriming or decapping pin removes the spent primer. During the process a new primer is press-fitted into the cartridge case, the case mouth is flared to accept a new bullet, powder is added to the cartridge case, and finally the projectile is pressed into the cartridge case and crimped in place. You can load a bullet faster than I can describe the process, depending on the gear you have.

As with shooting, driving, and using a chain saw, there is the possibility of danger when handloading. It takes serious study and concentration, but handloading can become a serious money-saving hobby. Many shooters enjoy reloading for its own sake. If you pay attention to detail, handloading is a safe and enjoyable hobby.

The primary reason for handloading is to save money. You can load your own ammunition for a fraction of the price of factory loads, largely because you are reusing the single-most expensive part of the cartridge: the cartridge case. Besides economy, there is always the fact that with care you will be able to create custom-grade ammunition that will equal the accuracy potential of the most expensive custom factory ammunition. You can tailor the load to the firearm and reach the perfect harmonic for accuracy. In some cases, you may even produce a load with more power than is the norm, given a particularly strong firearm. You may spend the same on ammunition as you do at present, but with the economy of handloading you will end up shooting much more.

There are a few things that you really need to have to hand-

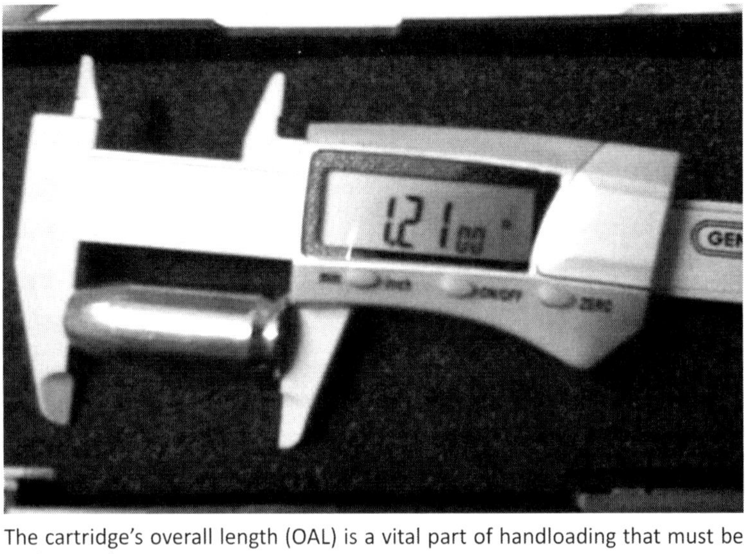

The cartridge's overall length (OAL) is a vital part of handloading that must be understood.

load. Starter kits by Lee, RCBS, and Hornady allow you to get started with most of the necessary items. These include a set of dies, the die press, a powder scale, and a powder measure. Most shooters begin loading with a single stage press. There is only one fixture that holds a single die at a time. The single stage requires that you change the dies between each stage. You will resize a bunch of cases, then change the die and flare the case mouth and prime the case, and finally charge the case with powder and seat the bullet. For slow production or occasional use, the single stage is fine, but this process gets old. The single stage is inexpensive, simple to use, rugged, and darned difficult to screw up—in short, an excellent press on which to learn loading. But if you have grown up using a loading press or spent time with a friend's and you know that you are going to be serious about reloading, then the smart money is on investing in a good three-stage press. I am pretty certain the very best of us may produce 60 to 70 rounds of ammunition on a single-stage press in an hour, but 50 rounds is more realistic. There are better means of loading ammunition.

The progressive loading press is best known as the Dillon. There are other brands, but none with the reputation of the Dillon. Its primary advantage is speed. A single pull of the handle performs several operations simultaneously. One hundred rounds an hour is leisurely production—some claim 250

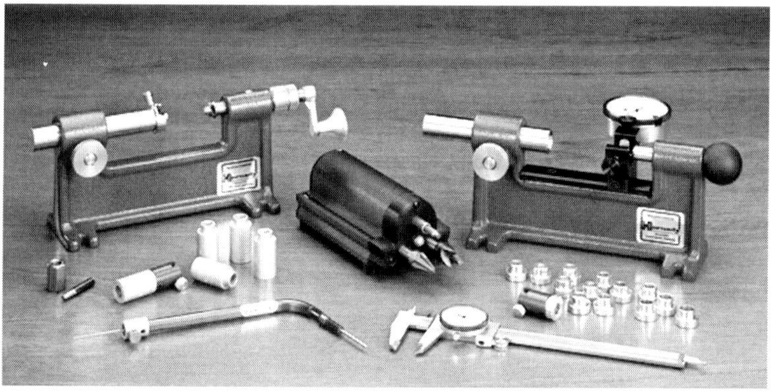

These case-trimming tools from Hornady are first class in every way.

rounds in an hour. The Dillon has some drawbacks you need to understand. It is fairly expensive but if you are going to be shooting a lot, the cost is amortized within the first year or so. There is also the issue of parts wear and breakage, but this consideration is outweighed by the Dillon's excellent warranty and customer service. Some lubrication and maintenance are required, which the multi-stage press does as well. There are a couple of presses I cannot recommend because they are so difficult to learn they are downright aggravating and they are simply too cheaply made. With the Dillon, your money is well spent.

If you are purchasing a single-stage, multi-stage, or turret press, then RCBS presses are difficult to beat. These presses are of excellent design. If you are loading rifle cartridges or foresee such use in the future, then nothing beats the legendary RCBS Rock Chucker press. The leverage of the Rock Chucker is needed for loading full-size rifle cartridges, such as the .30-06 Springfield. Manufacturer's recommendations will put you on the right path. Remember, there is nothing as desirable as a single-stage press to learn the craft, and the same dies may be used on the more advanced press. As an example, at the time of this writing, a Lee single-stage press runs less than $30. That isn't much of an investment in learning.

RCBS tools are among my favorites and will last for decades.

How do you choose the loading gear you need? Consider the amount of ammunition you will be using. Not the projected amount or the amount you hope to shoot, but a realistic assessment. If the volume of loading anticipated is 250 rounds a month or less than 1,000 per year, then the single-stage is probably adequate. Anything more than 3,000 rounds a year calls for a multistage press for economy of time and motion. If you are a serious shooter or competitor who is going to go to the range with a bucket of ammunition and fire thousands of rounds a year, then you really need to look hard at the progressive loading press.

The press is important, but the dies cannot be overlooked. They are the heart of the operation. The dies are manufactured for each cartridge and are specific to each caliber. For rifles, most are two-die sets while pistol-caliber reloading dies demand a three-die set. The operation is straightforward. With the rifle cartridges, the first die resizes the case and deprimes it as well. The case mouth will be flared or opened a bit to allow the insertion of a bullet in a tight press fit once the case is primed and the powder added. The second die is the crimping die. This die is used to properly seat the bullet and to crimp it in place. Crimping is simply forcing the walls of the case neck against the bullet in order to hold it in place. With the three-die pistol set, the operation is similar but slightly more involved. The first die resizes the cartridge case and deprimes the cartridge case. The second die, which will be mounted between the first and third dies in a three-stage press, opens the case mouth. This is usually the die that mounts the powder measure as well. The third component of the die set seats and crimps the projectile in the cartridge case.

When I began reloading, it was necessary to lubricate each case by hand or to roll the case on a pad as I ran it through the die set. To do otherwise was to risk having the cartridge case stuck in the die set, usually during resizing. Today modern carbide dies make lubrication unnecessary. Just the same, I often lubricate every tenth case or so as a precaution even when using carbide dies, and I also clean the dies from time to time. This ensures long life for the dies and prevents scarring the dies.

You will also need a shell holder that is specific to the caliber you are using, but these are a minor expense. Best to go ahead and buy the Lee set of shell holders, as they are inexpensive and the set will cover most of the common rifle and handgun calibers. You also need a powder measure of some type and a scale to double-check the measures. I have used the Lee powder scoops for so long and with such satisfaction that there is little need for anything else when attempting low-volume work with a single-stage press. They are usually spot on or very close to the listed grain weight, and they are useful for loading most of the common cartridges. There isn't much room for tweaking a load in tenth-of-a-grain increments as we often do for accuracy, but for the beginning loader on a budget, these scoops work well.

The more advanced gear includes powder scales that may be adjusted to throw any charge weight desired. You simply move the control dials or screws until the proper charge is confirmed by testing with a good powder scale. An RCBS Little Dandy measure is mounted on the Rock Chucker. I use an RCBS scale to check the weight every so often and tighten the adjustment screws every week or so. I would hate to be without my Lil' Dandy. This measure features changeable rotors to throw a pretty diverse array of powder charges. They are particularly well suited to use with pistol calibers on three-stage presses. Be certain you tighten the adjustment screws from time to time when using any measure, or you may inadvertently throw a dangerous charge. These measures are handy, capable, and well suited to most loading chores. The powder scale is used to confirm the powder charge. You cannot do without a good scale. The modern digital scale is expensive but well worth the price, but the various RCBS powder scales are accurate and durable.

You absolutely must have good loading data, and this doesn't mean anything scrounged from the Internet! (The Hogdgon powder on-line site is an exception to the rule, offering excellent information for loading cartridges.) Purchase the Lyman, Hornady, or Barnes manual—or better yet all three. Simply put, the powder manuals tell you what recipe to use. They outline the beginning charge for each weight of bullet. Changing a single component, such as the primer or the bullet, can change pressure even if the bullet is the same weight. Bullets are often longer or shorter from brand to brand; for example, a 160-grain .38 Special may seat deeper in the case with more or less of the nose exposed from one brand to the next. The deeper the bullet is seated, the more pressure. A long Keith-style semiwadcutter (SWC), as an example, features more of the bullet nose and weight exposed from the cartridge case, allowing a greater powder charge and usually more velocity with less pressure. A jacketed hollowpoint (JHP) of the same weight may seat deeper. The result is greater pressure for the same given powder charge. Do not carelessly change components without carefully consulting the loading manual. Always start with the recommended beginning powder charge—never jump to the maximum first. Cartridge cases and firearms differ concerning their tolerance for pressure.

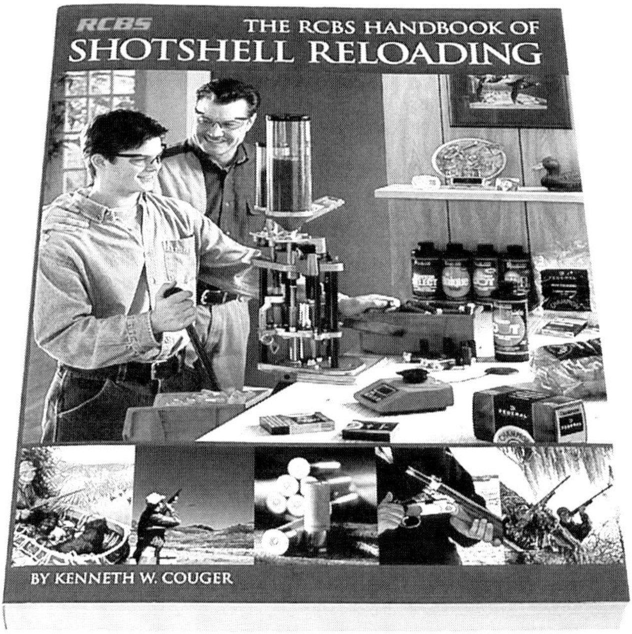

Get a good loading manual or, even better, get two or three. This one by RCBS is among the best.

A sobering thought—be careful with your loads and confirm your data! The case on the left, fired in a Kimber, was a tremendous overload. The ramped barrel of the Kimber saved the shooter from injury. The case on the right was fired in a Glock and burst. In both cases, loader error was to blame.

Among the great resources for loading data and one of the few on-line resources I can recommend is LoadData.com. *Rifle* and *Handloader* are both excellent magazines well worth a subscription, and LoadData.com is affiliated with both.

· · · CHAPTER 6 · · ·

WHAT MAKES A GOOD BULLET?

In this time of ever-increasing ammunition costs, how much are you shooting? Less than before? If you have the handloading thing down pat and you are turning out your own ammunition, you may be spending the same money but shooting more. Just the same, I don't think any of us shoot as much as we would like. While I use factory ammunition for consistency in many of my assignments, when I am shooting on my own time and my own dime, handloads with lead bullets are the rule. (I shoot primarily pistol rounds.)

As a handloader, I feel that I am more involved in my vocation and more knowledgeable about ballistics. A great understanding of how ammunition works and performs comes from making your own ammunition. As for economy, I think that the best point I can make to convince you to handload and use lead bullets is a simple one: we all have a finite budget.

You will spend about the same on your ammunition when you handload, but you will be able to shoot two to four times as many rounds of ammunition. That should be an incentive to begin handloading.

The primary advantage of lead bullets is that they are inexpensive, but there are other advantages as well. Lead bullets produce less wear on the rifling than jacketed bullets. A number of competitions actually require lead bullets. Older handguns, such as the Colt Single Action Army (SAA), require lead bullets for the best performance. For certain critical applications, we may prefer to use a premium jacketed bullet, but lead does the business for me about 99 percent of the time.

As for jacketed bullets, there is tremendous competition among the major makers. Each tries to convince the consumer that the latest jacketed soft-point bonded core bullet is better than the ones before. If there is a Hornady XTP bullet in the cartridge case, then you have made a good personal defense choice. The Hornady XTP is also a viable hunting bullet in many applications. Competitors include the Speer Gold Dot and Remington Golden Saber. These are fine bullets, but for 99 percent of the actual shots fired on the range, do you need a premium bullet? I agree that we should deploy only fresh, clean factory ammunition in defensive handguns, at least under normal conditions, but unless you have a Brinks truck full of cash following you around, you will never fire enough factory ammunition to attain real proficiency with the handgun. Lead bullets allow us to practice with what may seem a prodigious amount of ammunition to the average shooter.

Since we are cranking up the Rock Chucker Press primarily to conserve our ammunition budget, it only makes sense to use lead bullets. There are few range applications in which a lead bullet is not appropriate. There is little need to use jacketed bullets for most of our shooting. Just the same, lead bullets require an understanding of proper load practice. I am unwilling to admit that jacketed bullets are easier to load, except perhaps in automatic pistols. Lead bullets simply require some thought in loading when you are using the semiautomatic pistol cartridge.

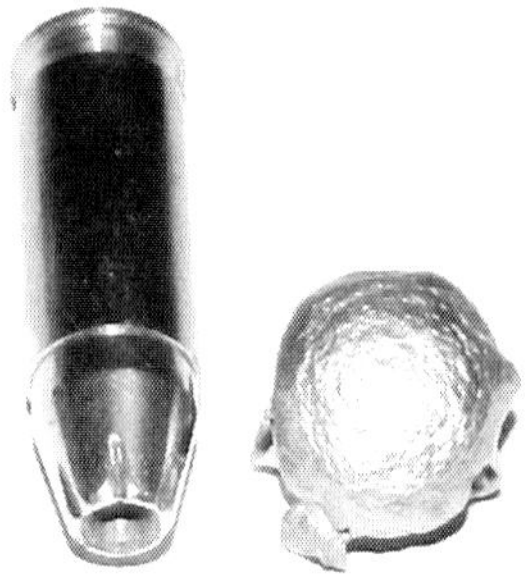

The Hornady XTP bullet is particularly well suited to high-velocity handgun loads, such as the .38 Super and the 10mm automatic.

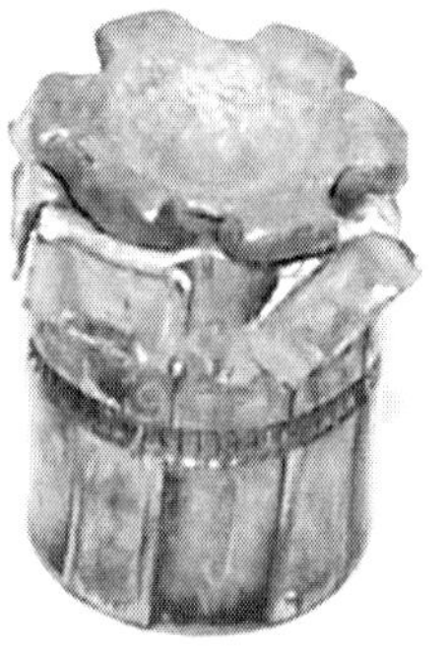

The author uses the Hornady XTP bullet extensively in handloading. This is a .38-caliber version, which has a good balance of expansion and penetration.

The .45 ACP cartridge works well with lead, jacketed, and plated bullets. These cartridges are loaded with Rainier plated bullets.

A common question concerning lead bullets is the proper bullet diameter. As an example, jacketed bullets in .45 ACP caliber are .451 inch and in .44 Magnum caliber .429 inch. Their respective lead counterparts are commonly .452 and .430 inch. There is good reason for the larger diameter. Lead bullets are softer with more give than jacketed bullets. This is good, as lead bullets will upset a bit to fill a bore. But the bullet must be larger in order to produce a good fit and gas seal. A bore-diameter lead bullet would be subject to gas leak. With gas leak, both accuracy and velocity suffer. Leading would be increased. While some leading is just fine and easy enough to clean, excessive leading creates a cleaning chore.

When embarking on the road to successful loading, some shooters are discouraged by their results with lead bullets. You simply must match the bullet diameter to your firearm. This isn't that difficult, as there are only slight variations among firearms, but for the best accuracy, the bullet must be matched to the particular firearm. An undersized bullet will give bad leading and poor results. The bullet that fits tight in the bore without much gas blowby is by far the best choice for minimal leading and good accuracy. The bullet doesn't have to be very hard for use in subsonic handgun calibers, such as the .38 Special and .45 ACP, but it needs to be hard enough to prevent leading if the cartridge is a high-velocity number, such as the .357 Magnum.

The bullet must be matched to the throat. The throat is in front of the cartridge chamber, and there is a taper from the throat to the forcing cone of a revolver. If the bullet doesn't fit well into the throat, then the hot powder gases will run all around the bullet, melt the base, and cause leading. Leading in the throat itself is from gas cutting. The bullet isn't the right size or perhaps the lube isn't that great. (I use the Red Rooster lube, the classic solution.) The bullet should be about .0005 inch less in diameter than the revolver throat. The bullet can be larger than nominal throat diameter if the bullet is fairly soft and the loads are low pressure. Sometimes poor accuracy is a result of a home caster attempting to cast bullets too hard. The

The .357 Magnum is among one of the most versatile, effective, and useful cartridges in existence. You do not need a 17-round magazine to be tactical!

Sometimes the smaller round is more expensive. The .327 Magnum isn't as great as the hype indicates, although it has its place. I find the .357 Magnum far more versatile.

bullet actually lets gas cut around the bullet because it doesn't plump up to fill the throat, resulting in both leading and poor accuracy. Some of the most accurate combinations in .45 Colt, as one example, come from Buffalo Bore ammunition, which employs a relatively soft alloy. These bullets are well suited to .45 Colt velocity. Revolver throats are sometimes oversize and eccentric, but the new Smith & Wesson revolvers, built on computer numerical control (CNC) machinery, are much better than anything offered in the past. This is particularly true in .44 and .45 calibers.

While there are many styles of revolver bullets, the very best for accuracy remains the properly cast Keith-style bullet. With a good bit of the bullet weight out front of the case and with a full caliber driving band, the Keith style is accurate and also effective on flesh and blood targets.

When loading for self-loading pistols, the differences in lead bullets must be understood. Autoloaders do not have throats and are also far less tolerant of less-than-perfect sizing and crimping during handloading. Some bullet styles are less likely to be tolerated on the feed ramp than is the case with the revolver. Just the same, there are excellent choices in the self-loading pistol that give fine results in both accuracy and effect on game animals.

The 1911-type handgun is often very consistent. As an example, I have used cast bullets sized .452 inch in practically every self-loader I have ever owned in this caliber (and also the .45 Auto Rim revolver). These bullets have uniformly given excellent results. The .45 ACP is pretty darned easy to work with compared to the 9mm Luger and often gives unexpectedly excellent results with a minimum of effort. I use a taper crimp and never a roll crimp in the .45, and it works for me. A few words are in order on leading. As long as the bullet alloy is sufficiently hard for the chore, leading will be minimal. In standard calibers, velocity may be kept under 900 feet per second (fps) and in Magnum handguns velocity may be kept under 1,100 fps for general use. A few hot-loaded rounds for hunting use will not be a problem. There are some alloys that are just fine at higher velocity, but a too-soft bullet will lead the bore quickly. Quality makers of hard-cast lead bullets—such as Oregon Trail, Leadhead, Midsouth, and Magnus—supply good lead bullets that do not lead excessively. Normal cleaning practice will keep the bore free of excess leading as long as the bore is cleaned every 500 rounds or so.

When you examine a commercial lead bullet you will note that there are grease grooves in the bullet base. These are for

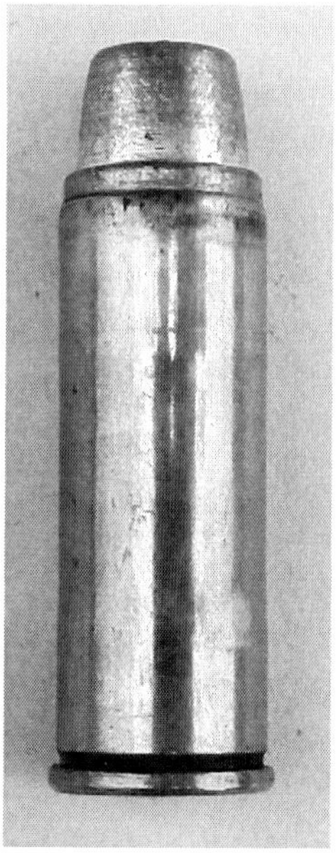

This pistol cartridge is loaded with an economical Oregon Trail lead bullet. The crimp is good and tight.

The Hornady XTP, left, is a great bullet but not necessarily for general practice. The Oregon Trail 200-grain semiwadcutter, right, is economical and accurate.

These hard-cast bullets from Missouri Bullet Company have given me excellent results.

Pay attention to hardness and proper consistency with cast bullets, and good results will follow. These are from Missouri Bullet.

These bullets from Missouri Bullet have proven accurate and clean shooting.

Starline Brass, Hodgdon Powder, Missouri hard-cast bullets, and careful assembly make for reliable accurate loads.

the bullet lube. Lead bullets require lubrication to avoid leading, especially at higher velocity. Various types of lubricant are advertised. I have used the Red Rooster brand with excellent results. Various makers of cast bullets use red, blue, and black lube of various compositions. They all work fine as far as shooting goes. A too-soft lubricant may melt and run into the powder charge, setting the shooter up for a misfire. The only realistic test I am aware of is to set a few bullets on the porch rails or a similar location in direct sunlight and see if the lubrication runs. You will note that most bullets designed for use in the standard calibers, such as the .44-40 and .45 ACP, feature a single grease groove. This is adequate for the velocity involved.

Elmer Keith designed his Keith-style semiwadcutters with two wide and deep grease grooves. Keith was among the first handgunners to work up loads exceeding 1,200 fps in a handgun. Conventional lead bullets would lead like mad. If too soft, they would occasionally skip through the rifling without actually taking a bite. The result was very poor accuracy and intense leading. The Keith-style bullet is ideal for use with high-velocity loads. These bullets also perform well on large game. They tend to drive straight, and the blunt nose produces a good wound cavity. These bullets tend to break bone rather than skid off bone.

A good barrel will often give fine results with lead bullets. I have a couple of handguns that have never had a jacketed bullet touch their bore, and I hope to keep it that way. It is true that a worn or pitted bore will not always give good accuracy with a cast bullet, but then the handgun will not often be accurate with a jacketed bullet either if the bore is in poor condition. Match-grade custom barrels, such as Bar-Sto Precision, in several of my .45-caliber 1911 pistols often show top-flight results with carefully loaded lead bullets.

A minor drawback—and I consider this a drawback of the handgun, not the bullet—is that lead bullets are not suitable for polygonal rifled handguns. Polygonal rifling produces a theoretical superior gas seal and may give a slight velocity increase

The polymer-frame Glock also features polygonal rifling, which isn't compatible with lead bullets. The Wilson Combat custom-grade barrel handles lead well with excellent accuracy.

This Wilson Combat barrel allows a Glock shooter to use lead bullets—and it also supports the case head in a superior manner.

over conventionally rifled barrels of the same length. While this may be true, I have installed a Bar-Sto Precision barrel in several Glock pistols in order to use lead bullets. Excellent results followed. In one case, in .357 SIG caliber, the Bar-Sto barrel produced not only greater accuracy, but also more velocity than the Glock barrel. Obviously Bar-Sto barrels are first class.

By purchasing a quality barrel of this type, you may use lead bullets in your Glock. I am certain that if you intend to keep the piece for several years, the investment is worthwhile, as lead bullet use is inexpensive. As a plus, the Bar-Sto barrel is more accurate with any type of jacketed bullet.

There are several types of lead bullets. The most expensive and least desirable are factory swaged bullets. They were used in several common factory loadings, and they are fine for noncritical applications. The 158-grain roundnose .38 Special load is perhaps the best known. Since this load only breaks 750 fps or so, the soft-swaged bullet is just fine. A bonus of this type is that many are hollow based. The extraordinarily accurate Winchester 246-grain roundnose lead (RNL) .44 Special is one example. This is another pedestrian load but one that is superior to any standard-caliber small bore. The hollow base expands under gas pressure and fills the grooves in the barrel well, producing excellent accuracy in good revolvers.

While ammunition for the big bores, such as this .44 Special revolver, is more expensive initially, you probably will not be burning up as much in practice.

Some shooters spend quite a bit of time having their revolvers' throats and leades reamed in search of accuracy with cast bullets when the use of a swaged bullet would produce good accuracy. Hornady offers several swaged bullets, including an accurate 250-grain bullet for the .45 Colt. If velocity is kept to 850 fps or less, the Hornady bullet with its white powder lube works fine. An advantage is that the Hornady .38 Special is offered in hollowpoint. This bullet expands at modest velocities and gives good accuracy. The problem with swaged bullets is that they lead more than hard-cast bullets. They are practically useless at velocities of more than 1,000 fps. Even a few rounds badly lead the barrel and adversely affect accuracy.

Just the same, when used in their proper niche, swage bullets can give outstanding results. One such example comes from Star Bullets. This 185-grain, semiwadcutter, hollowpoint (SWC/HP) has given excellent results in both .45 automatic and the .45 Auto Rim. Evidently Star uses a different composition than found in many swaged bullets, or perhaps the trick is in the lubricant. In any case, this bullet produces fine accuracy with minimum leading. While the hollowpoint is mainly for balance and for accuracy, keeping the nose forward and retaining a rear bias at velocities as low as 750 fps, the bullet shows some expansion. At 850 fps, it expands well. This is a bullet that will change your mind about swaged lead, at least in some applications.

The traditional hard-cast bullet is the product of a melting pot that dispenses hot lead into a mold. If you enjoy such things and have a well-ventilated area in which to cast, you will be ahead of the game if you cast your own. I sometimes use a Magma mold for my favorite .45-caliber 200-grain SWC. A melting pot and a sizer/lubricator are also needed. Today most of us will simply order the bullets from a major maker, such as Mastercast or Oregon Trail. These bullets are of high quality. While you may cast your own bullets for next to nothing, ordering bullets for certain calibers may be more economical. If you load primarily .45s, you may wish to order a few .38s or a number of .44-40 bullets, and these makers will accommodate your needs. I have enjoyed excellent results with these commercial-cast bullet makers. I normally order the standard dimensions, .358 for the .38 Special, .430 for the .44 Special, and .452 inch for the .45 ACP. I have enjoyed excellent results. The hard bullets catch the rifling well and deliver good accuracy.

OTHER USES

While I recommend against using handloads for personal defense, there are a couple of applications that beg for a custom load. I often load the Leadhead 200-grain flat point to about 850 fps from a 2-inch .38 Special. There is really nothing comparable from the factory, and this bullet gives ample penetration in this hideout handgun. In the .44 Special, I have used the AccuCast 250-grain SWC in my Model 21 Smith & Wesson with excellent results. Loaded to 850 fps, this is an accurate and reliable loading.

For hunting, nothing really matches a good hard-cast SWC. While there are heavier designs available, the Keith-style 250-grain SWC over enough #2400 powder to reach 1,400 fps remains a staple for hunting with the .44 Magnum revolver. As for pure accuracy, the Magna Cast 200-grain SWC delivers in my .45 automatic pistols. Lead bullets really do the business.

Since I mainly employ big-bore handguns, I am not overly tied to the hollowpoint bullet. A .44- or .45-caliber hole will do the business whether or not the bullet expands. I do admit a certain fascination with the .357 Magnum, however, and it is an important caliber around the house.

It is important to note that modern bullets are available that are a far cry from the inefficient roundnose lead bullet. The roundnose bullet was simply a transition from the round ball to a pointed bullet. It slips through flesh and does little damage. So a bullet that cut flesh rather than pushed it aside and did not bounce off bone was needed. Modern designs include well-designed semiwadcutter bullets that feature a flat point and sharp cutting shoulder. They will not be deflected by bone and will cut a full caliber hole combined with good penetration. There is also the ease with which light recoil loads may be put up with lead bullets. A jacketed hollowpoint bullet that is not driven at sufficient speed just may slip its jacket in the bore. With a good lead bullet load, you may work up a functioning load in the automatic that cycles just

Choose your bullets based on performance and avoid stocking up on second-rate performers. This is a CCI .22 Magnum loading.

The .22 Magnum is a little powerhouse that performs beyond its size. You should have one in your battery.

The .22 LR is a vital part of your ammunition cache. The CCI load is among the best when all things are considered.

fine but which is noticeably less powerful than factory loads. With the revolver, you can prepare powder-puff loads of around 600 fps or so that are a wonderful resource for breaking new shooters into centerfire shooting.

In today's tough times, I think it is safe to say that all of us are in search of greater economy. The lead bullet in general and the hard-cast lead bullet in particular deliver that economy. Performance in terms of accuracy and, if need be, penetration is there as well. Don't overlook this important resource. You will shoot more, and as you become more proficient, you will enjoy shooting all the more as well.

GAS-CHECKED LEAD BULLETS

Gas-checked lead bullets were designed to prevent gas deformation and leading in high-velocity loads. The gas check fits over the base of the bullet and works as advertised, but it is unnecessary for most loadings. During the sizing process, adding the gas check requires more time and a bit of finesse, so handloaders do not use a great deal of gas-checked bullets. My favorite is the Leadhead 200-grain flat point. Originally designed for long-range use in the .357 Magnum, I use the Leadhead sparingly in special-purpose loads in my snub .38. Penetration is excellent, and the bullet is quite accurate.

When it comes to lead bullets in rifles, we are often admonished that lead cannot be used in high-velocity rifle loads. This is true to an extent, but for inexpensive practice and small-game shooting, you are missing a lot by not using lead bullets. Oregon Trail offers excellent hard-cast bullets in 165- and 170-

grain weights for .30-caliber rifles. When velocity is kept below 1,500 fps, these bullets are excellent choices for practice. They are also effective small-game bullets, and a 165-grain .30 bullet, while very inexpensive compared to factory loads, is deadly on coyote and like-sized pests, including feral dogs.

While I like to keep a number of cartridge adaptors on hand, the experienced shooter and handloader may scoff at this recommendation and prefer to load up their own high-quality handloads with these effective rifle bullets. They are certainly worth your time to explore when cost is a consideration, as the economy is impressive in comparison to shooting jacketed factory ammunition.

JACKETED BULLETS

Jacketed bullets have their advantages. They sometimes are more consistent than lead bullets and often produce excellent accuracy. In the standard type of jacketed bullet, the cup is extruded from a machine, and the lead core is inserted. These are true jacketed bullets. Often they are among the most accurate, although in the revolver the properly cast lead bullet will often prove gilt-edged accurate.

Standard roundnose jacketed bullets are sometimes available in bulk at very attractive prices. If you are using a lot of ammunition in your high-velocity 9mm handgun, then purchasing a bulk amount of 9mm jacketed bullets, such as the affordable Winchester 115-grain full-metal jacketed (FMJ) bullet, may be a wise investment. They are easier to load and use than lead designs, and cleanup is considerably less bothersome. Often the price difference over lead isn't great if you purchase at least 1,500 jacketed bullets.

An advantage of the jacketed bullet is that a bullet may be designed that features a soft core suitable for expansion. Cover a hollow-nosed lead core with a thin gilding metal jacket and you may produce an excellent expanding bullet that will produce good upset even at pistol velocity. These bullets are often much more expensive than simple round-nose bullets and

Sadly, sometimes politics intrude into your decision about ammunition. If you live in one of the unhappy jurisdictions where hollowpoint bullets are illegal, Federal's expanding FMJ load is a great choice.

should be conserved for hunting or critical use. Just the same, often enough, jacketed hollowpoint (JHP) bullets are available at about the same price as FMJ bullets in bulk. The Winchester 115-grain JHP 9mm is an excellent design and is often priced as cheap or even cheaper than the FMJ version, largely due to mass production. I would take JHP bullets when possible because they are often more accurate for various reasons, including balance. The bullet just seems to be more accurate than a roundnose bullet.

One economical solution with few downsides is using plated bullets. These bullets are produced most famously by Rainier Bullets. Plated bullets are lead bullets with a flash plate of copper. They are excellent choices for economy and accuracy. A caution is that these bullets are often sized to lead bullet diameter as they are intended for use in bulk loadings and they are, after all, flash plated. So, be certain to use lead bullet loading data when working up loads for plated bullets. Also, when driven at higher velocity, plated hollowpoints may break up, but they do not expand in the normal manner that true JHPs do. That is fine, as they are accurate and intended for bulk shooting and economy. I have fired literally thousands of rounds loaded with Rainier bullets, and they have performed very well.

Plated bullets, such as these from Rainier, are a great economical resource for the handloader.

Plated bullets fall between lead and jacketed ones in terms of cost. Use lead bullet data when loading these bullets.

If you choose to load your own hunting or personal-defense loads, there are quite a few excellent bullets from which to choose. In handgun bullets there is nothing quite like the Hornady Extreme Terminal Performance, or XTP. Superbly accurate and offering an excellent balance of expansion and penetration, the XTP bullet is often recommended by experts as the ideal expanding pistol bullet.

Among the most popular service loads for peace officers are those using the Speer Gold Dot bullet. However, the inexpensive bulk bullets from Zero Bullet Company are decent performers and seem to expand well. If I were able to obtain a quality bullet, such as the Golden Saber, in bulk from Midway at a fair price, I would not sweat the small differences in JHP bullet performance. Differences do exist and the reader who follows closely will see that I have my favorites, but when preparing for the long term, economy is important. The real difference in effectiveness is where the bullet is placed. Just the same, the differences in the performance of these projectiles is worth studying.

It is important to realize that many plated bullets are not designed for personal defense, so cross these off the list. The XTP offers accuracy and a good balance of expansion and accuracy. After due consideration, the XTP is a strong favorite for meeting the needs of every situation and for always proving reliable and for failing to fragment. The XTP is sometimes available in bulk purchase and is not much more expensive than FMJ bullets. For bulk shooting and general plinking, the lead bullets are fine and often at least as accurate as a good jacketed bullet. But nothing will perform overall quite as well as the XTP.

In certain factory loads, Gold Dot offers proven performance. There is really nothing to equal the 135-grain Gold Dot .38 Special, and the 124-grain Gold Dot +P is a proven performer in 9mm Luger caliber. It is personal preference as to which JHP bullet to stock up on for the shortage, but frankly the better types are often in short supply. A few hundred of each should fill your needs for critical use for years.

In times of shortage, we are not going to be able to enjoy the luxury of working up loads at a leisurely pace as we do at present. We will have to carefully husband our resources. And the bottom line is that, even as I write these words, some of the expanding bullets we have relied on for years are difficult to impossible to obtain. As an example, the highly accurate Sierra pistol bullets are as scarce as hen's teeth at present. A premium JHP that is always match-grade accurate is the Nosler 185-grain JHP in .45 ACP caliber. This bullet opens up well at 1,000 fps or so, and it is as accurate as any hard-case lead SWC or premium jacketed bullet. I was able to open several 250-bullet boxes in the spring of 2012, but the 115-grain Nosler JHP in 9mm caliber has eluded me. These premium bullets are desirable and effective for accuracy and effect on target, but when it comes to burner bullets, there are cheaper choices that work just fine.

Jacketed Rifle Bullets

Jacketed rifle bullet technology is a science, but pretty simple bullets can work well. You must remember that when stocking up for the future, not to miss good buys that are available right now. There are arguments concerning the boattail hollowpoint (BTHP) and the jacketed softpoint. The boattail is a great idea but unnecessary inside 300 yards for most shooting situations. It may buck wind and turbulence better at extreme range, but there are other variables. If we are able to purchase a boattail bullet for the same price, then why not?

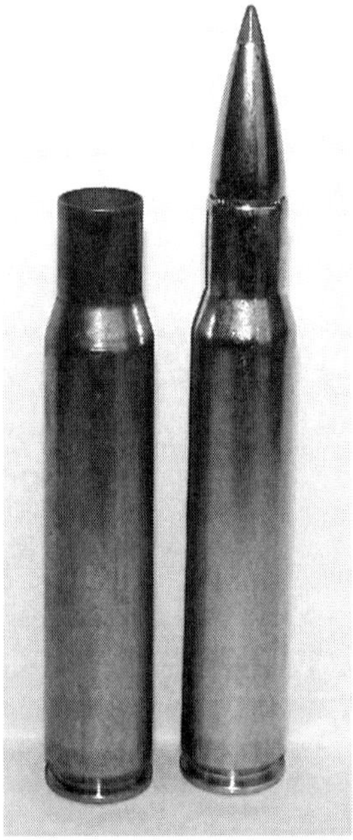

This rifle cartridge is loaded with the Hornady SST bullet. It is a good, tight fit in the case neck. This is a good choice for all-around hunting and personal defense.

The center of gravity of a bullet may be forward or to the rear, resulting in overstabilization or understabilization. A long hollow-nose bullet, such as the Sierra Match King (SMK) loaded by One Shot Munitions in match-grade loads, is a great bullet for long-range shooting. And while the 168-grain SMK is often touted for its accuracy, the 175-grain SMK is the better choice for extreme long range. This bullet stays straight and level in flight.

On the other hand, an expanding bullet suitable for personal defense or hunting needs to expand, and while the SMK is an open-nose bullet, it doesn't expand. When testing expanding bullets, don't get so tied up on weight retention. The only bullet that truly retains a large portion of its weight is the ballistic solid! Since expanding bullets do not need heavy jackets, they are made precisely and are truly versatile. My firing tests have actually shown flat-base bullets to be the most accurate. Remember, when you mess with angles, they must be perfect. The base cannot be bumped up and the bullet remain accurate.

Accuracy isn't always the long suit we are looking for. Frankly, with the level of accuracy possible with modern jacketed softpoint or jacketed open-nose bullets, any bullet from the major makers is going to deliver better accuracy than we could possibly hold.

Among the best deals and the best bullets for most of us is the Hornady Interlock. This bullet appears at first to be a conventional design with a cannelure and a raised edge on the in-

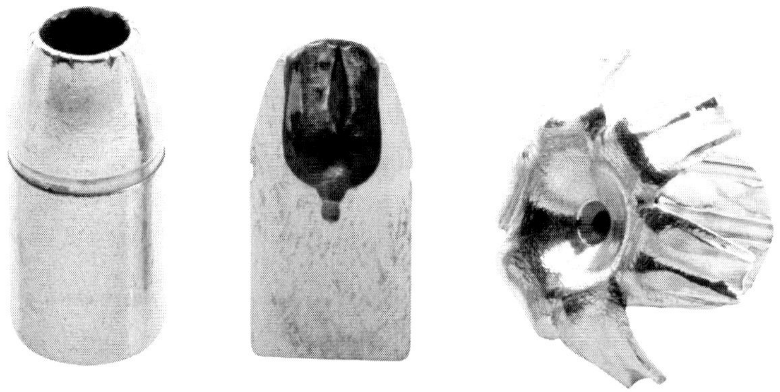

Economy isn't always the determining factor in bullet selection. Barnes all-copper bullets are more expensive than most, but they are also very effective.

side of the gilding metal jacket near the base of the bullet to prevent core slippage as the bullet expands—hence the name Interlock. It is simple and it works. The SST is another Hornady bullet featuring a red plastic insert in the nose that aids in expansion. The SST may expand more rapidly, but it is also an Interlock design.

I have two criteria for purchasing bullets I am actually going to use versus simply loading for shooting. They must be affordable, and, most important, they must offer excellent performance. Based on these two criteria, the Hornady SST is my top choice. If I am going to rely on a particular bullet and I am more or less married to it in that I will not be able to obtain any others, then the Hornady SST is the one on the shelf. It is the right bullet at the right price with simply excellent performance.

··· CHAPTER 7 ···

THE CARTRIDGE CASE

Today, brass is no longer the only choice for cartridge cases. Cases are manufactured from diverse materials, including aluminum, steel, and polymer. But brass remains by far the most successful material for cartridge cases.

Brass is malleable with a bit of give—it is strong enough to contain pressure but soft enough to expand without bursting. Straight-walled, tapered, or bottle-necked cartridge cases are easily formed from brass. But the greatest advantage to brass, within the context of this book, is that brass may be reused after each firing.

Brass isn't terribly complicated to manufacture. The primary element is copper with a smaller percentage of zinc. Copper alone simply isn't strong enough to withstand the sometimes high pressure generated by a cartridge, so some alloy is needed to harden the copper. The combination of copper and zinc has worked well enough for more than 150 years.

Note how smoothly these nickel-plated cases slide out of the cylinder of this compact .357 Magnum revolver. Ammunition quality is vital. Don't stock up on second-class loads.

Today there is a lot of focus on the perfect bullet. The American shooter demands it. Powders are not available that burn cleaner and produce more energy, but a great advantage to the experimenter is the fact that brass is stronger than either and more readily designed into new forms. Otherwise the short Magnum rifle cartridges never would have been possible.

As for brass cartridge cases, the alloy isn't complicated, and neither is the manufacturing process. A punch and die form the brass cartridge case from a sheet of alloy. The brass cup is formed, and then the final form is configured from other dies. The process that makes the brass stronger is called *annealing*. The consistency of modern brass is critical, and the major makers have perfected the science and the practical aspects. The heat-treating process is vital, as is the final consistency. The primer flash hole is stamped out during this process as well. I cannot overstate how important consistency is to the final product and to accuracy.

Probably the single-most respected cartridge case maker in the world today is Starline Brass. Many ammo makers, particularly such newer ones as Black Hills and Cor-Bon, rely on Starline Brass.

The primer pocket is an essential component of the brass cartridge case, but so is the case head. A generation ago, one of the leading causes of inaccuracy in 9mm pistols was the inconsistent case head. Not only was ammunition inconsistent in the world's most widely distributed pistol cartridge, but the pistols had to be able to feed, chamber, fire, and eject cartridges of varying specification. As a result, the chambers were less than match grade. Accuracy suffered. Today the situation is much better. The revolver cartridge will usually feature a large stepped case head with a case rim that allows the revolver cartridge to headspace on the case rim. Some early rifle cartridges, such as the .303 British and the 7.62x54 Russian, use a similar system. The more modern system is a case rim the same size as the case head with an extractor groove cut into the case rim. This system is used with self-loading pistols and must be consistent for reliability. Modern cartridge cases are usually very consistent.

The primary concern with brass is finding enough to reload. The first step is to scrounge the local ranges. Often public ranges have a bucket into which folks are required to drop brass. Others

My friend Ross donated these once-fired rifle cartridges to the cause. It is good to have shooting friends who do not reload but save brass.

let the brass lie on the ground, and it is fair game. Of course, it will be a diverse lot, but often you can trade for something of value. Never turn down any type of brass cartridge case, as you may just need it at a later date. At least, you may have the opportunity to trade it. I recall not long ago at all I was able to purchase 120 once-fired Winchester cases in .308 Winchester caliber. I was very glad to obtain these, as you could not borrow, beg, or buy brass in that caliber for months on end at the time. On another occasion my friend Ross gifted me with some 150 once-fired cartridge cases in .30-06 Springfield. These were mixed cases, but they served long and well once separated.

You can also obtain brass through commercial sources such as Starline Brass. This brass will be strong, consistent, and long lasting. It is more expensive than finding range brass or begging from friends, but the bottom line is this: Starline Brass is a reliable resource for quality brass and the brass will last for many loadings.

When firing pistol cartridges in bulk for personal-defense

practice at 10 to 25 yards, it doesn't pay to take the time to sort brass. For loading rifles or pistols for the finest accuracy, sorting brass is essential.

There is a certain amount of difference in the case capacity that can be measured with water. Naturally, the "tighter" cases that hold less water have thicker case walls. A lighter charge of powder—although the change may be very small, on the order of 1/10 to 3/10 of a grain—will produce the same velocity and pressure as a looser case. If you are loading high-end loads, such as full-power rifle cartridges, the difference can be significant. That is why we always start loading with a charge at least 10 percent below the stated maximum. Often you will reach a high level of efficiency before you reach the maximum. I have gotten stuck at a full grain below the maximum in pistol calibers and five grains short when loading rifle cartridges and maintained high efficiency and excellent performance.

These are a few cases that Alan Campbell has sorted and cataloged.

It is time to clean range brass. Dirt and grit will ruin dies if not cleaned from the cartridge case.

These cases have been carefully sorted and marked for caliber.

After being inspected for cracks, these cases were cleaned, resized, and de-primed. They will be loaded at a later date.

Shown is a bag of .45 ACP brass cases about to be processed and loaded.

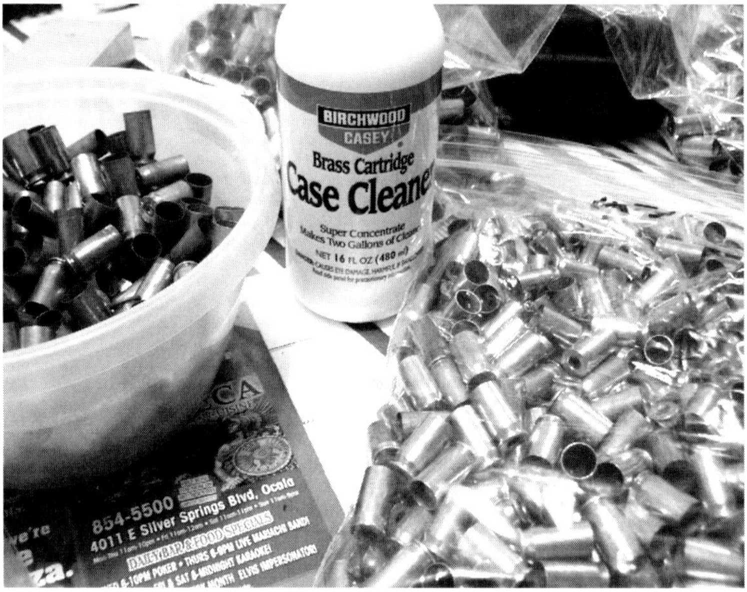

Birchwood Casey is among the leaders in case cleaning kits. They work just great.

Here are some dirty cases about to be cleaned.

The point is to sort your brass. While the pistol cartridges that show the least difference in accuracy are unsorted and loaded as is, it is never a bad idea to sort cartridge brass for the more important types of loads. Study the properties of the brass cartridge case, attempt to obtain quality brass, and learn to inspect brass for cracks and wear, and you will have a resource that is very economical.

CHAPTER 10

POWDER SELECTION

This chapter was written by Robert Alan Campbell.

Once you have studied the loading manual, you should have an understanding of how powder works and what type of charges you should be using for economy and performance. Simply put, the firearms cartridge is piston power, much like a car engine. Pressure builds when the cartridge ignites and pushes the bullet down the tube. The primary moment of highest pressure is when the projectile is first forced out of the cartridge case. Next, it is forced down the barrel and caused to rotate on its axis by the lands and grooves.

There is a significant difference in pressure peaks between fast-burning and slow-burning powder. A lot depends on how the bullet is loaded and whether it is deep-seated or seated to touch the lands (in a rifle cartridge), but in the end the powder used means much in pressure measuring. Most of what I load is pistol caliber. As a rule of thumb, the fast-burning powders are used for modest-velocity loads for practice and recreational

Powder selection is terribly important for both economy and performance. This is just a start if you have a battery of a half dozen or so calibers.

There are several ways of loading a case with powder. Here Alan Campbell is carefully weighing each powder charge for consistency. This is time consuming, but it produces great accuracy.

Carefully mark your loads. It is the only way to maintain control of safety and load development.

shooting. In most of the self-loading pistol cartridges, fast-burning powder is used in even the heaviest loads. The 9mm and .45, for example, are relatively compact cartridges that are very efficient with fast-burning powders. For hunting loads or maximum effort loads in the Magnum caliber revolvers, slow-burning powders are used.

There are tricks to choosing powder, and often the

Here are a few Magnum pistol cartridge cases and something to fill them with: Hodgdon Pistol Powder.

medium-burning powders are best for gilt-edged accuracy in revolvers. They sometimes generate modest pressure with a given charge and push rather than jolt the bullet. A tight crimp is an aid in a full-powder burn. A full burn simply means that all the powder is burning inside the barrel. When the powder burns outside the barrel, it is wasted and the result is a tremendous muzzle blast and flash.

When it comes to rifle powders, among the most versatile is Varget. The combination of *varmint* and *target* in the name tells a lot about the efficiency of the powder. I use medium-burning powders in the M1A1 Springfield and the M1 Garand because they are well suited to semiautomatic action and because they are close to the original specifications for the powder. I don't shoot Magnum rifle calibers, so I don't use slow-burning rifle powders. But if you begin to use powder charges much over 50 grains, you will find yourself going through a good number of 1-pound cans quickly. For this reason, I find the .308 Winchester with its relatively short efficient case about the largest practical rifle caliber in relation to the ammunition shortage. Brass is not particularly difficult to obtain in normal times, and .30 bullets are plentiful.

Today there are so many powder selections that it is almost funny to remember when I relied on Red Dot, Unique, and Bullseye almost exclusively. When I began loading Magnum revolvers, Hercules 2400 became a necessity. I primarily loaded for the .38 Special and the .45 ACP. Today, my loading bench includes dies for the .380 ACP, .38 Special, .38 ACP Super, .357 Magnum, .357 SIG, .40 Smith & Wesson, 10mm, .41 Magnum, .44 Special, .45 Auto Rim, .45 Colt, .45 ACP, .223 Remington, .308 Winchester, .30-06 Springfield, and probably a couple more. It would be easy to justify the use of more than a dozen powders for different uses, but we know that isn't practical and certainly not in the spirit of this book. While professionally I need to test many powders, on a personal basis, the fewer the better. When the handloading projects are not foremost, then only a few will get the job done.

There are always new and improved versions of old powders and carefully tweaked formulations. It is understandable that there is competition, and today I can state frankly that modern formulations produce higher velocity for a given pressure, run cleaner, and even produce better accuracy than ever before. The powder companies have given us wonder powders that burn cleanly—Titegroup is one example—but I doubt that even the greatest new rifle powders turn the .308 into a Mag-

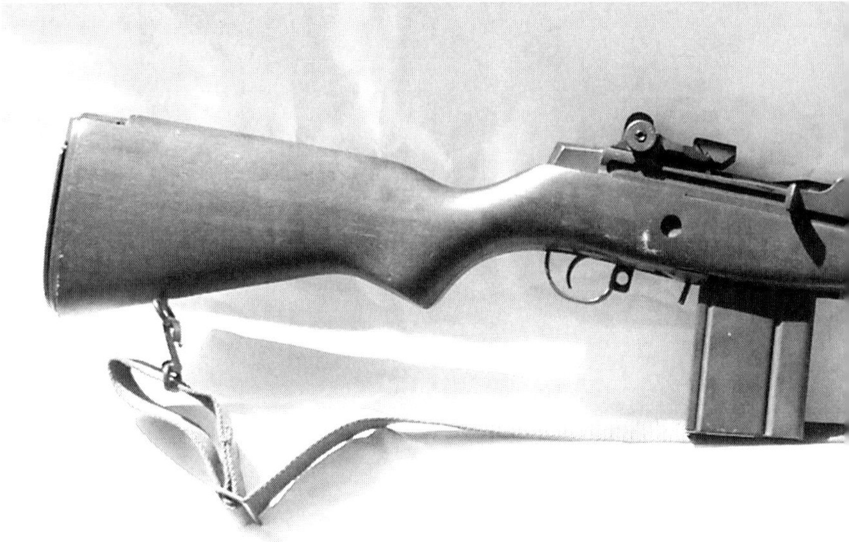

The Springfield M1A1 uses the powerful .308 Winchester cartridge. This is a very good heavy-duty rifle well worth its price. The .308 is widely distributed and relatively affordable.

num. We have to remember that the .308 Winchester, incidentally my favorite rifle cartridge, will do about 98 percent of what we really need to do with a rifle in North America. I fail to see how adding 75 fps materially improves performance, although I do see that increased firearms wear is a given with greater pressure. Even with equal pressure, greater velocity and momentum equate to greater wear on moving parts.

In short, the search for a great powder for every need can be expensive, and if we are interested in high-volume loading, we need to narrow the field in powder selection. Let's just look carefully at a wide array of powder and the results that the specific powder may turn in with, say, our main choices—in my case, the .223 Remington and the .308 Winchester.

If you live in the humid south or the frozen north, then you probably need a powder proven to be consistent across a wide range of temperatures. (This is mainly a concern with high-pressure rifle cartridges; handgun cartridges with less pres-

sure are not as great a concern unless you are loading Magnums right up to the edge.) You might also look at practical matters, such as how smoothly the particular powder meters. We do not wish to weigh every charge, although I do for accuracy loads. When beating the ammo shortage with high-volume loading, weighing every charge just isn't practical. I think that I could use Varget, the combination varmint and target powder, for the majority of my rifle shooting and not suffer at all. In handguns I could use Winchester 231 for just about everything except those Magnums. Hmm, sometimes the Magnums look like more trouble than they are worth, and a good standard-pressure cartridge may be the better choice. In many cases, I keep a single can of H110 loads for those times I need a true full-blown Magnum loading.

Someone may ask what the most economical powder is. As long as the powder gets the job done, then it is worthwhile and we have to look at the most bang for the buck. The chart that concludes this chapter illustrates the economy part of the equation. Since I load mostly for pistols, I focused on pistol calibers. Incidentally I have some pet loads covered as well.

Below is the list of powders with the caliber weight combinations. The asterisk indicates best performance from that powder (accuracy and grouping). As an example, in Magnum loads H110 seemed the best controlled in 140 grain—nice even blast but possibly more controllable. Some of the loads are experimental, but I pretty much have the .45s down pat. If I had to rank the powder by my personal favorites, it would be . . .

Pistol Powders

1. 231/HP-38
2. Titegroup
3. HS-6
4. Clays Universal (9/.38 applications)

Rifle Powders

1. Varget
2. H4895

What can I say of 231, HP-38, and Titegoup that hasn't already been said? HS-6 has *quickly* become one of my favorites. I would rank it up around the 231, but I am still trying it out for now. It is an uber-nice powder. Universal is better in the .38 and .357 for target loads with moderate recoil, and I found it very useful with a 147-grain cone for the 9mm. Trail Boss works wonderfully with lead in all but the 9mms: very nice, super-low recoil, and awesomely accurate. It is a nice low-target load.

Powder Price Formula: 1 grain= .000143 Lbs

Amount of grains per cartridge x .000143 = Pound weight

Pound weight x Price per pound of powder = Price per shell

Powder Pet Loads

* = Best Powder load in that caliber/powder combination

Titegroup

.45 ACP

200-grain RNFP 5.4gr @1.20
*200-grain LSWC 5.4gr @1.24
185-grain JHP Nosler 5.5gr @1.22
*185-grain SJHP PRVI 5.5 @1.20
*185-grain LSWC 5.5gr @1.24

.357

145-grain LSWC 5.2gr @1.57
*110-grain JHP Win Bullet 7.9gr @1.5
*110-grain SJHP Rem Bullet 7.9gr @1.5

.38 Special

*148-grain DEWC 3.3gr @(flush)
125-grain RNFP 3.8gr @1.35

9mm

126-grain LSWC 4.0gr @1.12
*115-grain JHP Nosler 4.6 @1.12
*115-grain LRN 4.2gr @1.10

Powder selection is critical for economy and performance. There is no better powder for the .45 ACP than Titegroup.

HS-6

.45 ACP

200-grain LSWC 8.3gr @1.24
*200-grain RNFP 8.3gr @1.20

.38 Special

125gr RNFP 6.5gr @1.35

9mm

*126gr LSWC 6.5gr @1.12
*90gr JHP Sierra 7.5gr @1.02

HP-38

.45 ACP

200-grain RNFP 5.5gr @1.20
200-grain LSWC 5.5gr @1.24

.357

125-grain JHP Win Bullet 8.0gr @1.5

.38 Special

125-grain RNFP 4.6gr @1.35

9mm

*126-grain LSWC 4.2gr @1.12
125gr LRN 4.2gr @1.12
*115-grain JHP Nosler 4.2gr @1.12

H110

.357

125-grain JHP Win Bullet 21.5gr @1.57
*140-grain JHP Sierra 15.5gr @1.58
158-grain SJHP Prvi 18.0gr @1.60

700-X

.45 ACP

*200-grain RNFP 5.4gr @1.20

9mm

*147-grain LFP 2.8gr @1.12

CLAYS

.45 ACP

200-grain LSWC 4.2gr @1.24

.38 Special

*125-grain RNFP 3.4gr @1.35

9mm

*126-grain LSWC 3.1gr @1.12

CLAYS UNIVERSAL

.45 ACP

200-grain LSWC 6.2gr @1.24

.357

125gr JHP Win Bullet 7.3gr @1.5

.38 Special

125-grain RNFP 4.5gr @1.35
*148-grain DEWC 3.8gr @(flush)

9mm

*147-grain LFP 3.2gr @1.12

LONGSHOT

.357

125-grain JHP Win Bullet 9.0gr @1.5

9mm

*126gr LSWC 5.7gr @1.12

W-231

.45 ACP

200-grain LSWC 5.5gr @1.24
200-grain RNFP 5.5gr @1.20
*200-grain JHP XTP 6.5 @1.23
*185-grain JHP Nosler 6.1gr @1.22

.357

110-grain JHP Win Bullet 8.8gr @1.5
110-grain SJHP Rem Bullet 8.8gr @1.5

.38 Special

110-grain JHP Win Bullet 8.0gr @1.5
148-grain DEWC 4.5 @(flush)
*125-grain RNFP 4.7gr @1.35

9mm

126-grain LSWC 4.2gr @1.12
125-grain LRN 4.2gr @1.12

TRAIL BOSS

.45 ACP

200-grain LSWC 5.2gr @1.24
200-grain RNFP 5.2gr @1.20

.38 Special

*125-grain RNFP 5.0gr @1.35

Note: Loads listed are for educational and economy purposes only. Use at your own risk.

• • • CHAPTER 9 • • •

PRIMERS

The primer is press-fitted into the primer pocket at the bottom of the cartridge case. When the firing pin strikes the primer, it causes the primer cup to be pressed against a small metal piece called the anvil. The priming compound is set off by shock and pressure from the firing pin. This creates a flame, which ignites the powder as it travels through the flash hole in the bottom of the cartridge case.

There are several different types of primers, including large pistol, small pistol, small rifle, large rifle, shotgun, and sometimes Magnum. Some—usually Magnum-type primers—have thicker cups than others, thus necessitating a harder firing pin strike. They must be harder to withstand Magnum pressure and not blow or be pierced by the firing pin. There are two primary sizes, .175 and .210 inch, with different compounds in the primer and different cup hardness. You need to remember

Alan Campbell is firing a .357 Magnum. Note the muzzle signature. It takes a hot primer to ensure a full burn.

to purchase the proper primer for each application. Primers have tripled in price over the last few years but remain affordable. For the best economy, purchase in 5,000-round boxes. I have used primarily CCI primers over the years with excellent results. Winchester primers are famously reliable as well.

While some claim that there are well-defined differences in primers, this is difficult to qualify in actual experimentation. It is true that sometimes changing a primer will produce minute differences in velocity, but as far as accuracy and reliability are concerned, the major makers—Federal, Speer, Remington, and Winchester—simply cannot afford to produce an inferior product. Primers were perfected more than 100 years ago. If you have a brand preference, then by all means stock up on your favorite.

Be certain to carefully plan your primer supplies. I fire pistol cartridges vs. rifle cartridges at a ratio of about 100 to 1, so I need to stock more pistol primers, both large and small sizes—say, about 5,000 a year. Conversely, a few hundred large

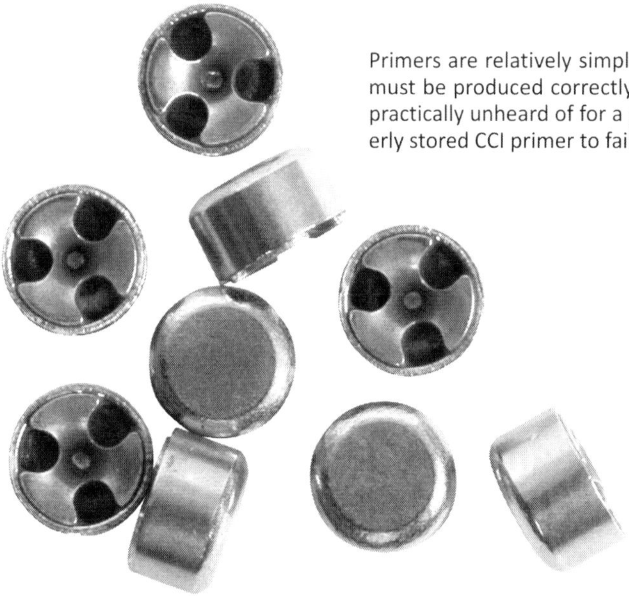

Primers are relatively simple but must be produced correctly. It is practically unheard of for a properly stored CCI primer to fail.

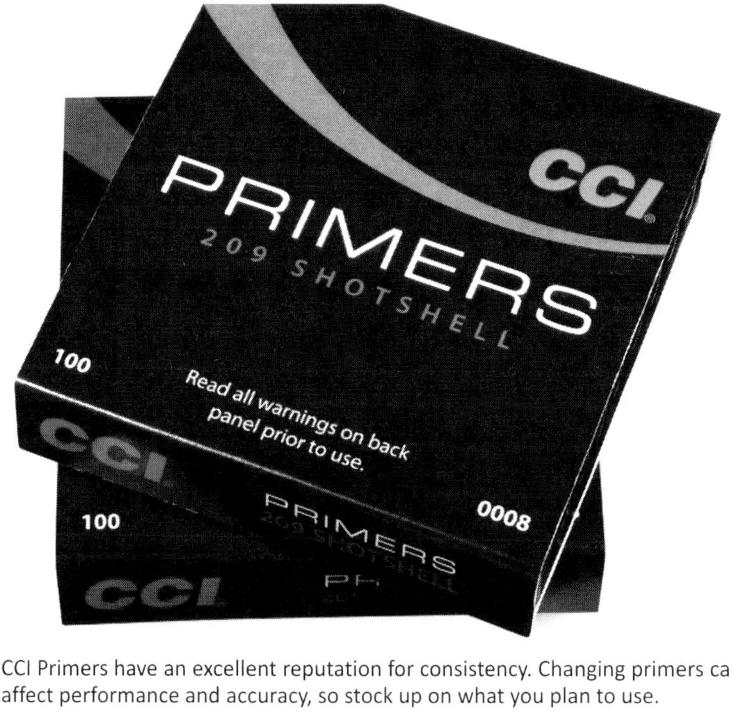

CCI Primers have an excellent reputation for consistency. Changing primers can affect performance and accuracy, so stock up on what you plan to use.

Even with an in-line muzzle loader, you need to stock up on primers. This single box might last a season or two.

primers for the rifle would do, and I seldom load the .223. The primers should be stored in a cool, dry place in their original containers. It is important that they be separated from each other, as it is possible they could detonate if dropped. This is why you leave them in their original boxes. Pouring a bunch of them in a shoe box would be a very bad idea. The few accidents I have learned of over the years have involved primers. Be careful in handling them.

Note: If you are reloading GI ammo, refer to the Hornady, Lyman, or RCBS reloading manual for instruction on how to deal with the "military crimp" on the primer.

HOW MUCH AMMUNITION DO YOU NEED?

Determining how much ammunition you need can be difficult. First of all, you need a cool, dry area in which to store it. Never put ammunition in a damp basement or a hot and humid attic. If properly treated from the factory with case mouth and primer seal, ammo is pretty resistant to the elements. But do not tempt fate, particularly if you are storing handloads. It is difficult to create handloads that are as resistant to the elements as the best factory ammunition, but a dab of nail polish on the primer goes a long way toward sealing the ammunition against the elements.

The primary concern for most of us these days is to have enough ammunition for an emergency. This means that you wish to have enough to survive the first few days of a civil disturbance. This is a serous consideration, well worth serious study. Civil disturbances can be frightening and deadly. I often laugh at the various TV movies that show folks pulling together in a crisis, whether it is civil unrest, a zombie attack, or a plague. Coincidentally, in the movies there is usually someone in the group with military experience. This supposed ex-soldier then makes plans in the middle of a square or other open area without regard for cover. Next he leads the mixed bag of citizens—whom he has trained in a day or so—into battle with no regard for cover or maneuvers.

Such events as the Northeast power blackout in 2003 and Hurricane Katrina in 2005 have shown clearly that during severe emergencies urban society will break down in near-tribal divisions. The scum of the nation use such breakdowns to settle old scores and as an excuse to loot, rape, and murder. In a natural disaster, help may materialize in a few days. If the disaster is more apocalyptic, then you need an adequate supply of ammunition. A lot depends on the predicted threat and your needs. To determine this, you must answer the following:

- What is your plan?
- Will you spray-and-pray a magazine full of .223 at every threat?
- How good a shot are you?
- Can you quickly dispatch a mad dog at 50 paces with a single shot?
- Can you take a deer cleanly at 125 yards with a single bullet?
- Do you intend to arm the entire family and provide each member with ammunition?
- How large is the extended family?
- Are you purchasing ammunition to ensure against the great what if, to barter in the coming emergency, or simply to make sure that your shooting sport remains affordable?

If you count the ammunition used for hunting in the past few years, I doubt you will have used much more than a hundred rounds. But we certainly do not wish to curtail practice and sighting in the rifle. If you are a handgun hunter, you realize that marksmanship is demanding and handgun skills are far more perishable than rifle marksmanship skills. The handgunner must practice in order to be formidable. The rifleman has an easier time maintaining useful skills; useful rifle skills may be maintained with a few hundred rounds a year. Of course, this does not apply to sniper-level skills, which demand thousands of rounds a year.

I recommend putting aside at least a four-year supply of ammunition now for the future and then maintaining the cache by reloading or buying what you are able to find. If you are a competitor and plan on competing in three-gun matches, then you will need to have on hand at least 10,000 rounds if you wish to remain competitive. Buying this much ammo amounts to serious money, and the fact is the ammunition shortage may prove too expensive for some of us. We may have to adjust our shooting lifestyle or lower our expectations. This is a bitter pill to swallow for some of us.

You need to think clearly about what you need and what you want. If you are going to be a miser and keep on hand only what you need for personal defense or food procurement, then a few thousand rounds should do. If you want to maintain your shooting skills through rigorous practice, you need 10,000 rounds for five years. Like others, you may find that once the ammo is stored, you are reluctant to fire it—after all, this is a major investment. You need to determine whether you are going to use what you store.

The next step is to initiate a plan for proper storage. The primary concern is rust and dampness. A rusted or corroded round is dangerous to fire and should be discarded. You have to control humidity in your storage area. The typical cardboard and Styrofoam ammunition box is fine for short-term storage of three to five years, but for longer storage, other types of containers are needed. As with all supplies, the rule of first in/first out applies. Fire the oldest rounds first and preserve the newer ammunition for the last use.

Ammunition stored in tins is the best bet. Fiocchi's Canned Heat line is a first-class means of storing ammunition without resorting to ammunition cans. I have used Greek military sur-

If there is any system better for long-term storage or a better buy than the Fiocchi system, I have never seen it.

Lapua Brass, CCI Primers—these are first-class loads stored the right way.

plus .30-06 in my personal Garand rifle that was at least 40 years old. It appeared as new when broken out of the gas, and my Beretta-made Garand rifle proved accurate and reliable with this ammo. 7.632x39mm ammunition is often available in similar tins.

If you are on a budget—and who isn't these days?—choosing a quality AK-47 rifle and plenty of ammo might be a great choice. But you need to determine your priorities before you purchase. With that in mind, choosing an AR-15 rifle in .204 Ruger or some other way-cool caliber may not be wise for the long term. Much the same goes for a 6.8mm rifle. The 6.8mm cartridge does hit harder than the 5.56mm, and there are good reasons for some to choose this caliber. If you do get a 6.8 rifle, be certain to obtain a

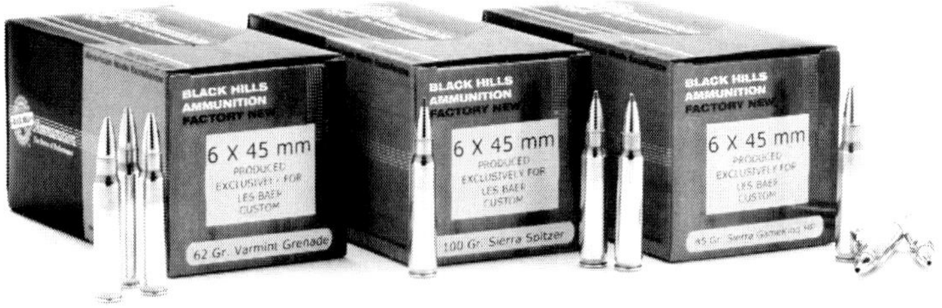

Black Hills 6x45mm cartridge has more energy and more wound potential than the 5.56mm NATO and solves a lot of problems. But you better stock up on Black Hills ammunition because it isn't available at the local sundry.

This is a storeroom at Black Hills Ammunition. It doesn't take long for these rounds to move out, so stock up now.

newer version with the 1:11-inch barrel twist rather than the older 1:10-inch twist rate. Pressure problems existed with early versions. With some loads, the latest style, sometimes called Spec II, results in better accuracy with the 110-grain bullets. But the 6.8 cartridge isn't widely distributed, and a rifle without ammunition is simply a club! Choose a rifle that chambers readily available ammunition. This would mean a 5.56mm over a 6.8mm and a .308 Winchester over the 7mm-08 Remington, despite whatever other advantages oddball calibers might seem to have.

I am familiar with the advantages of the .41 Magnum and

The .41 Magnum is a fine revolver and very useful, but it is neither versatile nor easy to find cartridges for it.

the .44-40 Winchester, as well as the .264 Winchester and the .300 short Magnums. You may feel you are playing it smart to have several hundred cartridges in your favorite caliber, and to an extent this is true, but what if the gun is lost or damaged or simply wears out? Then you have a stock of oddball ammunition and you may have a devil of a time obtaining another firearm in that caliber. If, on the other hand, you have a decent cache of .357 Magnum or .30-06, the problem will be lessened considerably. When planning for the great what if, you must think things through. In short, the cartridge you choose as your main focus should be a common and easily obtained one, as must the firearms that shoot it.

As far as storage goes, you must make sure there is no humidity trapped in the container when the ammunition is stored, which isn't that difficult. A compact dehumidifier will do the trick if the container is left often. Desiccant packs will absorb moisture. Once everything seems to be in line as far as storage goes, get hold of several surplus steel ammunition

These top-tier loads have been carefully marked for occasional use and testing the most accurate handguns.

cans. They are more expensive than they used to be but remain an affordable aid in storing ammunition. You should check the condition of the ammunition every five to eight months. Be certain to mark each ammo can carefully so that you will know the caliber and load inside. This is particularly true if you are storing handloads.

··· CHAPTER 11 ···

FACTORY AMMUNITION

PISTOL CARTRIDGES

.22 Long Rifle

The .22 pistol is an essential part of the battery for training and recreational purposes, but for personal defense the .22 pistol is a poor choice.

.38 Special/.357 Magnum

This is the single-most versatile combination available to pistol shooters. While there are limiting factors to the revolver, these calibers give great versatility; in the case of the .357 Magnum, there is no better defense cartridge. I include them both in the same section because every .357 Magnum revolver will chamber and fire the .38 Special cartridge efficiently.

From the lightweight Taurus 605 five-shot snub-nosed revolver to the heavy-barrel Smith & Wesson Model 27, the heavy .357 Magnum cartridge is well served by the understudy .38 Special. Of course, the .38 came first and quickly earned a spotty record as a manstopper. With the original 158-gain RNL bullet at 800 fps, the Special was seriously deficient. Today the .38 is still a fine practice and small-game cartridge, but heavy loads with well-designed bullets make the .38 Special a serious personal-defense cartridge.

With a highly developed load, such as the Cor-Bon 110-grain DPX, the snub-nosed .38 offers a degree of authority. This caliber is the best pocket-gun caliber and the lightest cartridge I am comfortable with for personal defense. With the 148-grain wadcutter, many 4-inch-barrel .38 Special revolvers are match-grade accurate and capable of taking small game cleanly. The .357 Magnum may offer twice the energy of the .38 Special and also make for a capable game cartridge for medium-size game at modest ranges.

The hot-loaded .38 Special is among the heaviest loads untrained shooters can handle easily. This is an important consideration if you are planning on family members with little or no training having access to the firearm. For the trained shooter, there are a number of full-power .357 Magnum loads that have no peer in the personal-defense realm. A personal favorite is the Cor-Bon 110-grain JHP, with a superior balance of penetration and expansion. For use when animals are a consideration, the Winchester Silvertip is a proven cartridge. This 145-grain JHP is flash-coated silver for looks but the performance is gilt-edged.

If you fire the Magnum a great deal, early wear is predicted. The revolvers won't blow up—the momentum is simply hard on small parts. If you plan on using the full-power Buffalo Bore hunting loads, for example, you may wish to invest in a Smith & Wesson Model 27 or a Ruger GP100. These rugged revolvers will digest thousands of Magnums before they need a tune-up. On the other hand, a steady diet of the heaviest Magnum loads will quickly wear a medium-frame revolver, such as the Smith & Wesson Model 66. I don't care if Granddad's revolver is still going strong, hard use can turn these revolvers into a rattling wreck over time and hundreds of rounds. Buffalo Bore offers a lighter Tactical load that, while impressive for personal defense, is not in the heavy-load category.

As you can see, there are several different levels of power inherent in the .357 Magnum and these levels must be matched to the shooter's ability. For longevity, a ratio of ten .38s for every Magnum round is recommended.

A good 4-inch-barrel .357 Magnum revolver is among the most versatile choices for anyone for personal defense, hunting, and general shooting.

9mm Luger

The 9mm Luger is the most popular and widespread handgun caliber in the world. It is not the most powerful or the most accurate, but the 9mm has several excellent points. It is very efficient. There is no cartridge the size of the 9mm as powerful, and the 9mm packs more punch than quite a few larger cartridges. The cartridge is available in pocket pistols to service small-size pistols. Some of the service pistols hold 17 rounds or more of ammunition, which is an impressive ammo reserve. The 9mm cartridge is compact enough that a great number may be stored in a relatively compact space.

On the negative side, the 9mm does offer uncomfortable recoil even in truly compact pistols, but the major problem with the 9mm is wound ballistics. The FMJ bullet simply does not have sufficient wound potential to perform well against living targets. The 9mm has the greatest deviation in effect among loads of any common handgun cartridge. Even with JHP bullets, my tests confirm that the balance of expansion and penetration is 100 percent different between loads ranging from 90 to 147 grains.

With a properly designed hollowpoint bullet driven to more than 1,200 fps, the 9mm is a formidable defense cartridge. The Cor-Bon 115-grain JHP is among the most widely distributed high-velocity 9mm loads, but in my opinion the newer Cor-Bon DPX loading offers the superior balance of expansion and penetration.

Another excellent choice is the Black Hills EXP, a fast JHP of 115 grains that does not break the +P pressure limit. The Black Hills 124-grain JHP +P is remarkably well suited to both pistols and carbines. Only careful load selection will give the user confidence in the 9mm, and these loads are certainly capable.

Confidence in your equipment and proficiency with the chosen gear are vital, and the 9mm can be mastered and used well. Recoil is light enough and follow-up shots are predictably accurate. In the end based purely on economy, storage, and overall efficiency, the 9mm is difficult to argue against. If you stockpile plenty of inexpensive FMJ loads for practice and a smaller amount of high-performance loads for defense use, the 9mm will serve you well. Currently my son is deployed by the U.S. Army and carries his Beretta 92 9mm on a daily basis. He is well armed, both in terms of his firearm and his mental preparation. The thinking man is the real weapon.

This World War II Remington Rand has proven perfectly reliable with all Black Hills loads, including the 185-grain JHP.

.45 ACP

The .45 ACP is my favorite handgun cartridge. While some cartridges may be viewed as "good enough," the .45 ACP is a proven manstopper and among a very few handgun cartridges that do not rely on an expanding bullet for effect on the target. The .45 ACP cuts a .451-inch hole through the target. It will take deer-size game at modest ranges, although it is not ideal for this work. The advantage of the .45 is that it is the most powerful cartridge well suited to a relatively compact self-loading pistol. The advantages of having more shots than a revolver and relatively controllable recoil are evident in the .45 ACP.

Ammunition is bulkier than some cartridges, but the .45 will do more work than lesser cartridges. The .45 ACP is a low-pressure cartridge. This means the life of the weapon firing the cartridge is predictably longer with less breakage. The .45 ACP is chambered in the 1911 handgun, a pistol with unequaled ergonomics and speed to an accurate first shot. If you prefer the SIG or Glock pistol, then there are .45 ACP versions of these readily available as well. If you handload, the .45 ACP is an easy cartridge to load to good accuracy. The .45 is easier to fire with lead bullets than most of the other self-loading cartridges.

Among the better and more consistent loadings for the .45 is the Black Hills 230-grain JHP. This load demonstrates a good balance of expansion and penetration while maintaining excellent accuracy. For those who prefer a compact pistol, the choices are good. The 160-grain Cor-Bon DPX is a specialty load designed for use in 3-inch barrel .45s. I could go on, but the fact is this cartridge is remarkably consistent and offers an excellent set of ballistic properties.

A final word on the .45—if you are primarily a handgunner and the handgun will be your focus, as in a rural environment, then the .45 is the top choice. If you are primarily a rifle shooter and regard the pistol as a backup for occasional use, then a revolver or a 9mm pistol might be the better choice. The .45 isn't the easiest combination to master, but it is an efficient one.

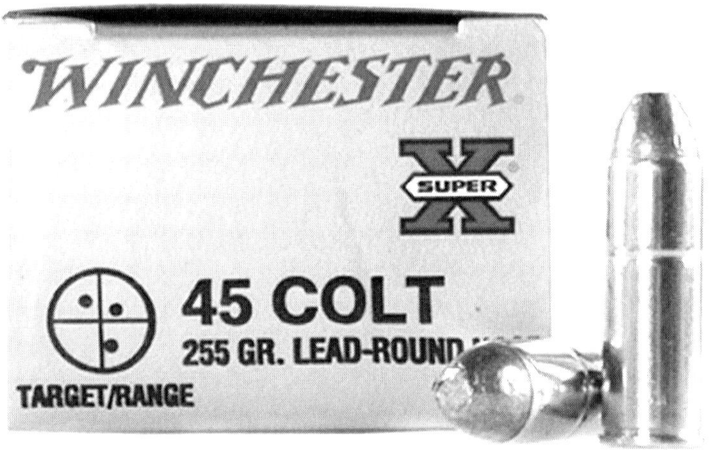

Winchester's .45 Colt load is among the most desirable antipersonnel loads. It is a bit pricey, so load your own and count on Winchester for serious use in old cowboy pistols.

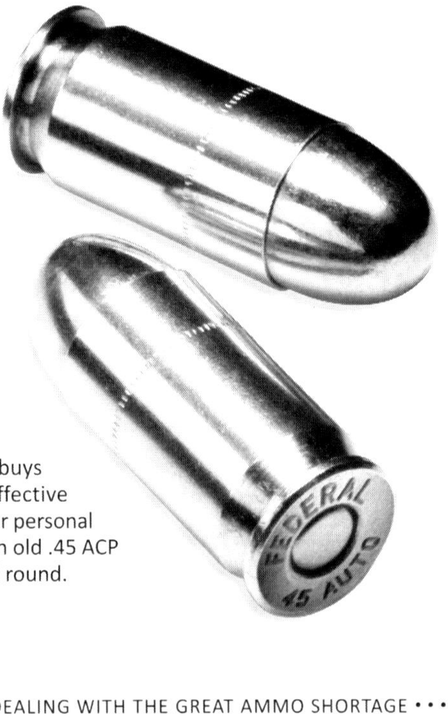

Among the best buys and the most effective handgun rounds for personal defense is the plain old .45 ACP full metal jacketed round.

Remanufactured Ammunition

When discussing the ammunition shortage, I don't mean just storing ammunition. You need to consider how to survive on a day-to-day basis and how to feed your handgun for regular service. In this regard, police agencies have relied on remanufactured loads for decades. The civilian shooter is less aware of this resource. There are different means of obtaining remanufactured loads. For example, Black Hills ammunition offers a first-class product using fired brass. The company does not ask for trade-in brass; it simply obtains used brass from commercial sources. Other concerns ask for a trade-in or discount trade in brass. In the end the quality of the product doesn't depend as much on the brass as the care taken in manufacture.

··· CHAPTER 12 ···

SURPLUS AMMUNITION

When it comes to ammunition, too often myths and misinformation are an impediment to advancement in training. Training properly demands lots of ammunition. It should be stressed that the two things you do not wish to run out of in a fight are time and ammunition. The same is true in training. Invest time and ammunition in training and train properly. For this, affordable ammunition is important.

During the lifetime of the firearm, a serious shooter will spend more on ammunition than on the initial purchase of the weapon. Everyone wants clean-burning, accurate ammunition. One of the best resources for good-quality affordable ammunition is military surplus sales. All of us are painfully aware of the increasingly high costs of components and ammunition. Several of the major makers are now offering steel-cased ammunition, a sure sign of a problem in obtaining raw material.

There are times when quality surplus ammunition is available for about the same price as putting together personal handloads, but without the time and labor. Surplus ammunition was often produced at a time when prices were lower, and the savings were passed on to the user.

While handloading is rewarding on many levels, the time spent in loading the prodigious amounts of rifle ammunition used in training and the occasional competition is a daunting proposition. For occasional shooters or hunters, high ammunition cost is more of an inconvenience than a hardship. Firing a modest amount of ammunition to sight the rifle is all they look for. More is better for marksmanship. Action shooting is a different creature than casual shooting. Often the cost of ammunition is the most significant factor in a match. When looking at surplus ammunition, do not attempt to understand global markets and wonder why a country is offering this ammunition. Enjoy the bargain. The U.S. military is using all the 5.56mm, .308, and 9mm it can obtain, and so are our allies. Just the same, there is ammo available for those who are willing to search diligently.

Many shooters shy away from corrosive primed ammunition, and a significant quantity of the available loads are corrosive primed. These shooters have heard of corrosive primers turning the bore red (or rather the residue from such primers), rusted bores and flash hiders, rusted bolts, and generally corroded firearms. And all overnight! Handloaders have cursed crimped-in military primers, which make for added difficulty in reloading corrosive primed brass. You also must clean the brass more thoroughly to remove all corrosive residue before reloading. Sure, you have heard it all, and there are still plenty of shooters using surplus brass and ammunition. Not only do they use this ammunition for training for the most part, but many also trust it for personal defense and competition shooting.

For the better part of the 20th century, corrosive ammunition was in use by all the world's armed services. Practically every round of small-arms ammunition fired in World War I and the majority in World War II were corrosive-primed

This surplus ammunition features excellent primer seal, but it is corrosive primed.

rounds. The first move to noncorrosive priming came with the introduction of the .30 M1 carbine. The military's first low-maintenance rifle, the carbine required noncorrosive ammunition because maintaining the carbine did not include scrubbing the gas cylinder as was required by the M1 Garand. This move to lower maintenance eventually resulted in the gas-impingement system of the AR-15 rifle, although the M14 rifle demanded more care for reliability. While the United States began the move from corrosive ammunition during the 1950s, many nations continued to produce corrosive-primed ammunition until at least the 1960s. Match-grade and sniper-grade rounds were corrosive primed, so obviously this ammunition was not second class compared to modern loads. It was simply primed differently.

What is corrosive priming? Corrosive primers contain

potassium chloride and sodium chloride, both very stable. These primers are completely stable at both very high and very low temperatures. Soldiers operating in the Aleutians or the sub-Saharan desert can be confident these primers will always fire and that they will be consistent and not demonstrate pressure spikes. While modern technology no longer demands these compounds, in their day they were useful. Corrosive primers were a slight inconvenience compared to the previous black powder military cartridges.

Potassium chloride and sodium chloride are actually not corrosive on their own, but rather they are considered corrosive because they attract moisture. If they were corrosive themselves, they would rust out of the primer during storage. The bolt, barrel, bore, action, gas cylinder, and any muzzle brake or flash suppressor are especially susceptible to corrosion brought about by these compositions. In short, corrosive primers have no negative effect on reliability but rather enhance reliability. The conditions under which the loads were stored is more important than the priming.

Surplus loadings are produced around the world, and while accuracy and quality control differ, European surplus is often very good. Canadian 9mm Luger ammunition was once highly prized because it would run a German Luger with good function and accuracy, while domestic loads would not. Greek .30-06 ammunition was highly prized for reliability and accuracy. Yugoslavian 8mm is a good tool for assessing the true accuracy potential of an original Model 98 Mauser. This military spire-point-bullet-loaded ammunition is more accurate than commercial loadings and considerably stronger as well. Thousands of rounds of com-bloc 7.62x39mm, a smaller quantity of 7.62x54mm, and even a small lot of Russian 7.62x25mm with the 1943 head stamp have been fired on the local range without a single burp. This stuff was old but hot and accurate, which proves that the Russians sometimes got it right.

Just how accurate is surplus ammunition? Military ammunition is loaded with reliability uppermost. That is why the corrosive primers and crimped-in primers—and sometimes

I don't issue a blanket recommendation for surplus ammunition, but the cartridges illustrated here were accurate and clean burning.

heavy-bullet crimps as well—were used. Mass production with ammunition that must always stay within allowable pressure is also important. As a rule of thumb, good-quality surplus will be as accurate as commercial generic ball loads and often more accurate than foreign-produced economy loads, and then everything in between is possible. The results will be commensurate with the firing platform.

A recent test with a Bushmaster carbine exhibited groups of 2 to 3 inches at 100 yards with just about any ammunition. Inexplicably one of the three Bushmasters on hand is noticeably more accurate than the others, but the edge isn't there with average ball ammunition. It is best demonstrated with Black Hills A Max, which usually groups three shots into a little over an inch to one-half inch with this standard carbine. Most surplus is as accurate as most generic ball, as a rule of thumb, and more accurate than the dirty foreign commercial stuff. It will not be as

If you fire corrosive ammunition in your firearm, there is only one sure way to clean it—hot water!

accurate as the Black Hills 60-grain JSP or the Black Hills 77-grain Open Tip, but then what is? Surplus loads are accurate enough for practice and all but benchrest competition.

I should make the point that not all surplus is corrosive, but the majority is. You should regard all ex-military ammunition as corrosive primed and follow the proper cleaning procedure unless you definitely know otherwise. There is no easy way to tell if the ammunition is corrosive primed.

Now we come to the dirty part of the story. Cleaning the firearms is a necessary chore, which some shooters seem to think they can neglect with modern ammunition. If you're unwilling to clean the ammo, you should avoid corrosive-primed loads. In fact, you should avoid shooting much at all because every firearm needs to be cleaned often to prevent eccentric wear from powder, lead, and jacket material buildup. To clean a firearm that has been exposed to corrosive priming, you follow a program similar to that used for black-powder guns.

The old "run it under the hot water trick" certainly is an option for cleaning, but one that should be used sparingly. Rather, the program begins with carefully fieldstripping the piece (but not completely disassembling it) and then using powerful ammonia to clean the bolt, bore, trigger mechanism, and, in those firearms that use them, the flash hider and gas cylinder. Clean the parts thoroughly. Some use simple household glass cleaner, which also works great. Occasionally, you should run the parts under hot water after cleaning with ammonia. It is not necessary to use boiling water. Many much-used firearms are a testimony to the effectiveness of this program.

When all is said and done, surplus ammunition is a good resource for those who wish to shoot plenty of ammunition at a fair price. Thousands of shooters have used surplus ammunition for many years and have enjoyed excellent service. That is all we can ask of anything manmade.

At this point I should mention that Hornady offers ammunition in a number of very difficult calibers, including the .455 Webley, 8x57mm Mauser, and .303 British. As a result, the need to locate surplus ammunition in obscure calibers is largely alleviated. While Hornady's loads for obsolete calibers are more expensive than the general run of surplus ammunition, it is affordable, clean-burning, and often loaded with a modern expanding bullet suitable for hunting.

Note: Be very careful when dealing with bulk or repacked older surplus that looks like it may have been tumble cleaned. The problem comes not from the occasional misfire, but rather from the fact that when loaded cartridges are tumbled (in walnut shells or whatever), this tumbling/vibratory motion tends to break some powders into smaller particles, which increases its burning rate and leads to higher-than-normal chamber pressures.

··· CHAPTER 13 ···

OUT OF THE BLUE

As the ammunition shortage waxes and wanes in some areas and becomes more acute in others, makers have mobilized to take advantage of the situation and offer products for ammunition-hungry consumers. Some offer good products, and others get by as cheaply as possible. Some of the ammunition isn't reliable for service or personal defense, and some isn't consistent and clean burning.

WOLF PERFORMANCE AMMUNITION

A new face on the ammo market has made quite a reputation in the United States, Wolf Performance Ammunition, also known as WPA. One of the leaders in the inexpensive ammunition market, Wolf is noted for its use of steel-cased cartridges. While as inexpensive as most surplus ammunition, Wolf is non-

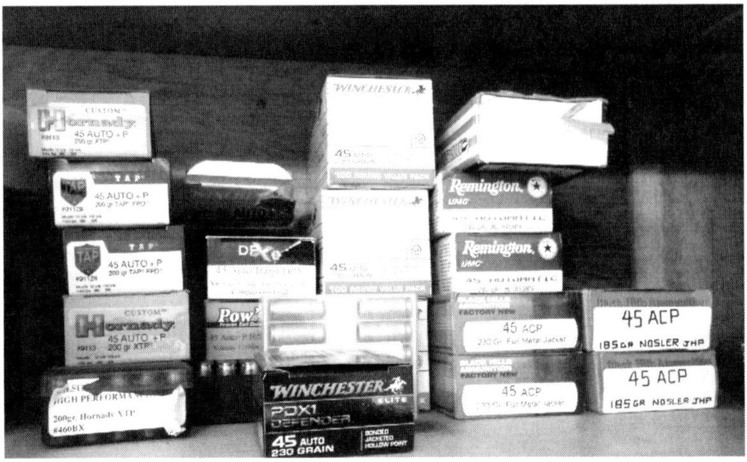

This shooter has followed the catch-as-catch can approach and purchased whatever he could find at a fair price, including a wide mix of brands. It is all good ammunition, with a couple of specialty loads.

This shooter has followed a more systematic approach by purchasing as much Black Hills 9mm as he could find. This is not a bad approach by any means.

This is just a small fraction of the supplies needed to keep an interested shooter happy for a year or so.

When the chips are down, the ammunition has to be reliable. Consider this when choosing personal-defense loads. This home defender loads his MechTech CCU conversion of the Glock 19 with Hornady 124-grain XTPs.

corrosive primed. For many years Wolf used lacquer-coated cases, primarily to avoid corrosion, but it switched to polymer-coated cases, which both prevents corrosion and slips into the chamber more easily. My experience with Wolf has been positive. Its loads have given good performance for the money, with excellent reliability. Although the cartridges are not particularly accurate in match-grade rifles—say, in .308—they are more than sufficient for those firing the AK-47 rifle. Pistol cartridges are accurate enough for training and reliable. In 8mm Mauser and .303 British, recent tests showed Wolf at least as accurate as any other commercial loading, making the Wolf loads a good choice compared to expensive domestic loads.

Wolf has branched out into hard-to-find calibers, such as

This is a trick setup and a very good one. The MechTech Systems CCU is on the Glock 19 frame. With quality ammunition, such as the Black Hills 124-grain +P, this combination is accurate to 100 yards or even a little more.

the .303 British and the 8mm Mauser. This allows owners of these rifles to bring them out of the closet and actually fire and enjoy the old warhorses. Wolf has also introduced modern loads in the once very expensive 6.5 Grendel. A drawback sometimes noted with steel-cased ammunition is that the steel case does not adhere to the chamber as tightly, resulting in more carbon or unburned powder. I also believe that this powder accumulates in rifles using Wolf ammunition simply because Wolf is inexpensive and shooters often fire prodigious amounts of ammunition during one range session. This is more of a consideration than a drawback since it can be easily remedied with a few more minutes spent cleaning the firearms. Steel cases are definitely harder on the extractor than brass cases,

Surplus ammunition is usually reliable, but much depends on storage. As a rule, I prefer Wolf ammunition for use in older rifles.

but this matters little in the AK-47 or AR-15, which are designed for hard use. In the end, steel-cased cartridges may increase extractor wear by a few percentage points. Today some Wolf is brass, and some is polymer coated.

Wolf is a good resource for inexpensive bulk ammunition. It will not be the most accurate and not always a good performer with expanding bullet loads, but in my experience, Wolf is reliable.

BROWN BEAR AND SILVER BEAR AMMUNITION

I have used a good amount of both Brown Bear and Sliver Bear ammunition, primarily from DKG Trading. This ammunition has given good results. In particular, the pistol caliber using hollowpoint bullets seems to open up reliably. I have also fired a 7.62x39mm hollowpoint load, and its performance was impressive. This would be a good all-around defense load, but penetration is not adequate for deer-sized game. This load has proven difficult to find during the past year or so and may be unavailable now. The Silver Bear ammo is a little smoother in operation and a bit more expensive than Brown Bear, but both are accurate and reliable.

These Russian makers—Brown Bear, Silver Bear, and Wolf—have a place in the scheme of things with affordable and reliable ammunition. There have also been a number of regional companies offering ammunition for sale, often at gun shows. During the previous ammunition shortage, many were asking scalper prices. I cannot comment on quality, as it is impossible to test and gauge a representative sample of all these makers. I do know that much of the ammunition wasn't useful.

I have tested a number of loads from one concern with a bullet that beyond question was a great design of cutting-edge technology. The first box of ammunition performed well, while the second had serious issues. Some rounds would not chamber at all, while others would chamber but were so tight the slide could not be racked to clear a round. The crimp was obviously not correct in those.

When you intend to store large amounts of ammunition for future use, it is best to purchase from a company you know has a good reputation for quality control, as well as the reasonable expectation of remaining in business. Many of the fly-by-night concerns have poor reputations for quality.

FIOCCHI

Fiocchi is a relatively new name in the ammo business, but it has a strong presence in the American market. Fiocchi has been making ammunition since the dawn of smokeless powder in the late 19th century. It is among the most important suppliers of Italian military and police forces and has been so since World War I. Fiocchi is not strictly a foreign company, as Fiocchi America has a manufacturing plant in Ozark, Missouri.

I have to stress that Fiocchi is not low-budget ammunition,

This young lady is burning up a target courtesy of Fiocchi .22s and Zombie Targets Inc.

although it is often available at an affordable price. Fiocchi has earned a reputation for performance and accuracy, and offers ammunition that is good enough to bet your life on in the pistol calibers and rifle calibers that are well suited to hunting game. The shogun loads are a particular standout, with a great number of loads that are well thought out and appropriate for any shotgun sport. Among the finest shotgun slug loads I have ever fired are the Aereo Slugs in 12 gauge. These slugs give good accuracy and offer excellent effect on target.

Many of the personal-defense loads in the pistol calibers use the Hornady XTP bullet, an excellent choice. If you do have a pocket pistol in .25, .32, or .380, Fiocchi is usually loaded hotter than domestic loads in the FMJ line. This ensures their functioning in European pistols. Fiocchi is an excellent resource for shooters.

... CHAPTER 14 ...

VERSATILITY

In this chapter I will examine the best choices for all-around use. There are a few cartridges that are jack-of-all-trades and that will serve well. One of these is the .22 Long Rifle. Perhaps the best means we have of beating the ammunition shortage is to use the humble and affordable .22. Let's look hard at the question of rimfires for training.

RIMFIRE TRAINING FIREARMS—GOOD ECONOMY OR WASTED BRASS?

A proven resource in creating a marksman is the use of inexpensive .22-caliber firearms and ammunition. The rimfire offers little or no recoil, minimal report, and good accuracy. It is recognized that the rimfire is a good training aid for pure marksmanship; that is, trigger control and learning sight alignment and sight picture. The rimfire is not necessarily a good choice for tactical training. Yet in today's tight economy we see both .22-caliber conversions and dedicated .22-caliber firearms pressed into service in training.

With the high and increasing costs of training, the .22-caliber conversion appears to be a good buy at the first look. However, we must fully understand the system, including its advantages and drawbacks. First, the use of .22-caliber ammunition is economical. Any way you slice it, the difference in price between rimfire and centerfire ammunition allows the shooter to fire 10 rounds of rimfire for the price of a single service cartridge. Plus, the .22 offers practically no recoil, and it may be used in ranges that would not be safe with high-powered firearms. But there are drawbacks as well, and these must be recognized.

On the debit side, often as not police recruits have little firearms experience. Few of today's recruits grew up shooting. While the recruits may be trained quickly and do not have to unlearn bad habits instilled from poor instruction, their unfamiliarity with the .22 is a shortcoming. When these shooters purchase a .22 for practice, it is often their first experience with rimfire firearms. They have already been indoctrinated into the centerfire firearm. The .22 isn't a miniature centerfire—far from it—and some of these shooters are disappointed.

Let's look at the construction of a .22-caliber rimfire cartridge. The bullet is of soft lead and is heel based, meaning simply that the bullet is the same diameter as the cartridge case. A heel or rebated rim on the base of the bullet allows the bullet to be pinched into the cartridge case. This construction is not as robust as that of the centerfire, which typically features a jacketed bullet crimped into the cartridge case. The rimfire gets its name from the type of priming compound. The .22 Long Rifle is the last survivor of the rimfire cartridges, primarily because it is not used for critical duty. The service-size .38 and .44 rimfire cartridges have long been obsolete. The inside of the cartridge case is primed around the cartridge case rim. The .22 headspaces on this rim. When the firing pin strikes the case rim and crushes it against the firing chamber, the priming compound creates a flame that ignites the powder in the cartridge case. This type of ignition is not as reliable as centerfire technology. Rimfire cartridges are reliable in general but not as reliable as the more modern centerfire cartridge.

Due to the rimfire construction, .22-caliber firearms may be damaged if they are dry-fired, as the firing pin will strike the firing chamber. The heel-based bullet is also not as sturdy as the crimped in centerfire. (The more modern .22 Rimfire Magnum uses a crimped, jacketed bullet.) The bullet is more easily bent away from the cartridge case during cycling. As an example, when the rimfire fails to properly feed due to a bullet shifting in the case, an immediate action drill ramming the bolt forward often makes the jam worse.

The powder used in rimfire cartridges is intended to maximize the cartridge in a rifle barrel. This powder often fails to burn completely in a short handgun barrel. As a result, the .22 is a dirtier loading than any properly loaded centerfire cartridge, and powder ash builds up more quickly.

All these drawbacks are apparent to sportsmen and shooters who grew up shooting the .22. Tactical shooters practicing with the rimfire may be disappointed when they find that the .22 begins to malfunction around the 300-round mark. Cleaning is needed more often. The bullet is less robust than a centerfire in handling. When firing thousands of rounds, there will often be several misfeeds and even a failure to fire the inside primed rimfire cartridge. Rimfires may prove unsettling to a shooter who knows that the centerfire Beretta, Glock, or SIG will go thousands of rounds without any sort of malfunction or without cleaning.

The value of the rimfire is unquestioned but those who adopt the .22 for training must understand the nature of the cartridge. I have dealt with many shooters in training who are disappointed with the performance of their .22-caliber firearms. While quality is sometimes an issue, even with the best-quality rimfire firearms, perfect reliability is not in the cards.

A question that often comes up is whether the shooter

should purchase a dedicated .22-caliber firearm or a quality .22-caliber conversion unit. Various .22 conversions, such as the Ciener pistol conversion, have been available for some time. While the units vary in quality, there are good examples available for the 1911-type handgun, SIG, CZ, Beretta, and most popular service pistols. There are conversions available for the AR-15 and Ruger Mini-14 rifle. These conversions have many tactical advantages, so their use is ideal if there is a conversion available for your service firearm. If not, there are other considerations. For example, a young officer on a strict budget may purchase a .22 and leave it at home for the spouse to defend the homestead. You may wear the service pistol while practicing with a dedicated .22. There are always tradeoffs to be considered.

TRAINING CONSIDERATIONS

We have discussed the characteristics of the .22-caliber cartridge. As stated, it must be cleaned more frequently, and it is less reliable. In addition, the "advantages" of the .22 that are often touted in training are not necessarily advantages in tactical training. For example, the light recoil and low muzzle blast are ideal for marksmanship training since they do not distract from the fundamentals of marksmanship, which is fine for target shooting, but tactical shooting demands movement and firing quickly at moving targets.

These tactical drills are not the same as target shooting. The primary mistake most shooters make when using rimfire firearms for practice is failing to firmly grasp the firearm as they would when using a centerfire firearm. They relax the grip because there is no recoil. They breeze through training and ace the course with minimal effort. The cadence of fire is also disrupted. When firing the service pistol, as one example, the cadence of fire is not set by how quickly you are able to press the trigger but by how quickly you are able to reacquire the sights and control recoil. The .22 does not present the same set of problems. Shooters may find themselves firing much more quickly and perhaps making five hits with the .22 in the time it takes to make two good hits with the .40-caliber pistol. This is not necessarily a good thing. If this training crosses over to the centerfire handgun, the shooter will be firing too quickly and missing. In short, the shooter using the .22-caliber firearm for tactical training must use the same strong grip and hold he uses with the centerfire, and he should time his shooting in the same manner as with the centerfire: fire, control recoil, reacquire the sights, and fire again. Only in this manner will rimfire training cross over to the centerfire in a profitable manner.

When it comes to rimfire training, the distance is not as important with handguns. Rimfire or centerfire will keep your training within 25 yards. With the rimfire rifle we may use reduced range targets to make training relevant. Pepper poppers and other steel reaction targets must be set for the .22 rimfire, or they will fail to register hits. Often even a light ping isn't heard despite a direct hit with the .22.

There are many considerations, but as you see there is much that must be done correctly in order to make rimfire training profitable. Rimfire training will never take the place of live fire with the service gun and ammunition, but it can be a cost-effective option for training. With a few simple considerations, a rimfire-training program will be both profitable for the shooter's proficiency and light on the shooter's wallet.

DUAL CALIBERS

We have discussed the advantage of the .38 Special and .357 Magnum combination, which is truly the most versatile all-around handgun-ammo combo. No self-loader would allow you to carry a shot load first up followed by two Glaser Safety Slugs and then three full-house Magnums! But there are a few semiautomatic handguns that allow the use of several calibers.

Among the most common is the .357 SIG/.40-caliber setup. By simply changing the barrel, either cartridge may be fired in the handgun. I am not a huge fan of either, but I admit the .40 has a good reputation in personal defense. Contrary to praises in the popular press, the .357 SIG has issues due to the jacketed

This well-worn old Ruger .22 features dual-chamber capability. It will fire the .22 Short, .22 Long, .22 Long Rifle, .22 Magnum, and shotshells in two calibers. That is versatility!

This shooter looks formidable. Her AR features a CMMG .22-caliber conversion and Black Dog magazine.

bullets in police service, which are primarily designed for penetration rather than dynamic expansion.

Another interesting combination is the 9mm/.38 Super. However, the 9mm +P+ now offers such excellent performance that the .38 Super is not quite as attractive as it once was. The .224 bottleneck on the 1911 frame is usually accompanied by a 9mm barrel for good reason. This isn't a fad likely to catch on.

Possibly the greatest single setup for a pistol is the .45 ACP 1911 handgun with a rimfire upper. This combination can be wonderfully accurate and afford many hours of low-cost practice. The CMMG conversion for the AR-15, coupled with Black Dog magazines, is first-class and well worth its modest price.

I am far less impressed with the concept of a pistol and a rifle firing the same cartridge. A rifle should chamber an effective cartridge. Even if the scenario calls for primarily urban use, the .223 rifle has every advantage over a 9mm or .45-caliber carbine. The .357 Magnum lever-action rifle is a great shooter with plenty of power for personal defense and has some merit, but the cartridge was created for handguns, not rifles. Still, this one remains an option for shooters who are comfortable with the lever-action rifle.

The Brooklyn Special—as some call a short, handy rifle dedicated to personal defense—is a fast-handling carbine for all-around use. My personal Rossi carbine is deadly accurate well past 100 yards with the Black Hills 158-grain JHP load and is a joy to fire and use. With this carbine and the Smith & Wesson Model 66, a shooter would be well armed, providing he was skilled in the use of these firearms. The diverse loads are an advantage. The need to have only one cartridge on hand is an advantage to some, and while a pistol caliber carbine isn't a rifle, the power of the cartridge is greatly enhanced not only by the full-powder burn in a longer barrel, but also by more accurate delivery from the long gun.

.223 Rifles

By far the most versatile single rifle for all-around use is the .223 Remington. A good AR-15 type rifle will defend the home, serve in the wildest civil disturbance, and bring home meat—although for larger animals, the .223 is pushing the envelope. For example, during the notorious 2011 incident in Zanesville, Ohio, in which the police were forced to down a number of dangerous exotic animals, including lions and tigers, that the owner of a wild-animal preserve had let loose, the .223 was effective only if 15 or more hits were made. At least the hits came easily and quickly from the easy-to-handle .223. If a bear is in the line of sight, then the .308 is definitely needed. But because the .223 is so light, easy to use, accurate, and reliable, it is difficult to fault. As far as a go-anywhere, do-anything rifle, the .223 is a great choice for most situations.

This Barnes solid bullet would be useful for defense against large animals. If a bear is part of the threat profile, then a few of these loads are necessary.

.308 Rifles

This ammunition is heavier and more difficult to store, and the rifles are heavier. Plus, lightweight rifles kick more than the .223. But you can do things with the .308 that are impossible with the .223. From the 125-grain Cor-Bon urban load to the 168-grain Hornady A-Max, there are .308 loads that simply perform like no small bore can. The .308 in a properly set-up rifle is highly accurate, and with the proper indoctrination, rapid follow-up shots when needed are not difficult. The .308 quite simply has much to recommend on the basis of performance.

CHAPTER 15

CARTRIDGE ADAPTORS

When it comes to the ammunition shortage, the more means we have of utilizing different calibers, the better. Over the years, neighbors, friends, and well-wishers have given me half-fired boxes of ammunition comprising calibers from .32 ACP to .454 Casull. It can be useful to have an adaptor to employ what ammo is on hand in a more powerful firearm.

Subcaliber adaptors are not the same as a .22-conversion unit. Caliber adaptors are substitute chambers that are inserted into the chamber of a centerfire rifle to allow a smaller cartridge to be fired in the rifle. This cartridge must have the same general bore diameter as the rifle. For example, the .32 ACP pistol cartridges uses a .308-inch bullet, the same as the .30-06 rifle. So, the smaller-caliber bullet will work just fine in the larger rifle if the proper subcaliber adaptor is used.

Cartridge adaptors are available in quite a few combinations. They offer real versatility in high-power rifles.

Simple to use and effective when used properly, these adaptors can be real money-savers.

It requires imagination to find practical applications for the adaptors, but in the end they are inexpensive enough to give them a try. An application that comes to mind is introducing young shooters to centerfire rifle shooting. In a .308 rifle with a subcaliber insert and the .32 ACP round, recoil is nonexistent and accuracy good enough for close-range use. You may even convert a centerfire rifle into a small-caliber game getter. I have examined quite a few adaptors, and the majority have good materials and machining. Even the ones supplied in unfinished or painted steel worked just fine. There are only so many variations, and most seem to work well. Accuracy varies with the firearm.

Let's look at the adaptor itself, which is specific to the firearm. In other words, you must have an adaptor in .308 Winchester to .32 ACP or .30-06 Springfield to .32 ACP. The adaptor resembles a cartridge case with a milled extractor groove and a bottleneck case. The bottom of the cartridge adaptor, which holds the subcaliber cartridge (such as a .32 Smith & Wesson Long to be fired in the .30-06), is removed and the subcaliber cartridge placed in. (You can use a spent case to remove a spent case from the adaptor.) If you wish to have more than one shot in a bolt-action rifle, then you need more than one adaptor. I have not done this. I have adaptors in .32 ACP, .32 Smith & Wesson Long, and .30 carbine in the .30-06. These light pistol cartridges do not have enough sectional density for accuracy past 50 yards, but up to that range they can be pretty accurate.

The .30-caliber carbine is pretty loud compared to the .32 ACP, but it is much more effective. In fact, this hotshot little

In testing, the adaptors worked just fine in a variety of firearms.

round makes your .30-06 rifle a great all-around varmint and coyote gun. The .32 ACP is more of a squirrel gun. The .32 Smith & Wesson Long was a dandy small-game round when handloaded with a 100-grain Magnus Cast Bullets semiwad-cutter to about 1,000 fps; however, if only factory loads are considered, use the .32 ACP. The .32 Magnum revolver round with the Black Hills 85-grain JHP would be a pretty neat sub-caliber adaptation as well.

When properly constructed, the subcaliber adaptor features a tight fit in the chamber. There is a minimal jump from the adaptor to the barrel leade and pretty decent accuracy. Some will deliver a 2-inch group for three shots at 50 yards. This is from a rifle capable of 1 1/2-inch groups at 100 yards with full-power .30-06 ammunition. The little .32 ACP Hydra-Shok and the Speer Gold Dot are capable in .32 ACP, as are other calibers. The .30 carbine I mentioned with the new Cor-Bon DPX is an excellent small-game and pest load. Accurate well past 100 yards, this is the preeminent .30-carbine load and well worth your attention.

Adaptors in .22 rimfire for use in the .223 Remington are a little more complicated. In this case, you remove the base of the adaptor, slip in the .22 cartridge, and press the base of the bullet back in. This allows the centerfire firing pin to strike the base of the adaptor and then proceed through a second firing pin, which strikes the rim of the rimfire cartridge. This arrangement would seem to increase lock time, but just the same, the conversion is valid; factory ammunition is accurate to 25 yards and perhaps a bit more with a quality bolt-action .223. I have used the Winchester Super X with good results in my AR-15 rifles. The little self-lubricated lead bullets do not lead the barrel much, if at all, and have proven capable of taking small game cleanly.

An odd situation exists with the small-caliber pistol rounds, such as the .32 ACP and .32 Smith & Wesson Long. As a rule, pistol-caliber carbines in 9mm and .357 Magnum pick up as much as 400 fps from a 16-inch carbine barrel. This means greater velocity, more bullet expansion, and sometimes more penetration and thus more killing power. But the .32 ACP actually loses velocity. The minute powder charge isn't enough to propel the bullet all the way through a long barrel and still produce high velocity. As an example, the Fiocchi 71-grain .32 ACP is among the fastest loads at some 1,000 fps from the typical pistol barrel. From the average bolt-action rifle, the load is achieving no more than 780 fps. Evidently resistance in the barrel is at play.

I did play with a heavy charge of slow-burning powder in the .32 Smith & Wesson Long, this time with the Hornady 60-grain XTP. While this load clocked a decent 1,000 fps in the 4-inch revolver (compared to about 650 fps for the factory 98-grain roundnose lead load) from the various rifles in which I clocked this load and adaptor combination, the best gain was perhaps 25 fps. Bottom line: you cannot count on gaining velocity with the small calibers.

Naturally, the .22 Long Rifle, .22 Magnum, and .30-caliber carbine rounds will develop normal rifle velocity. These little adaptors might just be the thing for those who wish to invest in as few firearms as possible but maintain as much versatility as possible. The adaptor will allow small-game hunting with a full-power rifle, as well as ample practice in areas where the report and penetration of a full-power rifle would be unwise.

••• CHAPTER 16 •••

FIREARMS SELECTION

HANDGUNS

It is always tempting to recommend your pet handgun, the most powerful or most accurate handgun, or the one that has won the great battles in your life. But the handgun is far more individual, even personal, than other types of firearms.

There are many factors that come into play when choosing a handgun. A handgun needs to be suited to your body size, level of training, and recoil tolerance. Most of us carry our handguns concealed, and a handgun that weighs more than 35 ounces is ill-advised for concealed carry. Some people may tolerate more weight, some less, but this is a rule supported by studies conducted by federal law enforcement agencies. If the handgun will only be kept at home ready or carried as a field gun, then a larger firearm is fine. No handgun is too big to fight with. For example,

It can be argued that a good .22 and .45 are all the handguns you need. I agree that is true for 99 percent of shooting chores.

It would be wise to take careful notes and evaluate performance before investing in a truckload of shells or ammunition.

Among the finest choices for all-around practice and shooting enjoyment is the Ruger .22/45. A good .22 is essential to your battery.

the 6-inch-barrel Ruger GP100 maximizes the .357 Magnum cartridge, is very accurate, and is among the most controllable revolvers in rapid fire. But the piece is too large and heavy to count on having at your side at all times. If there is an ammunition shortage, chances are things are a bit dicey and you will want to have your handgun with you at all times. By the same token, a handgun that is too small and too light is difficult to control. A short sight radius is difficult to use accurately. Larger handguns are more powerful and often more reliable as well. The chances of small parts breaking is greater on smaller handguns, as small parts are more stressed in recoil. When choosing a handgun, you have to balance comfort, accuracy, and controllability.

There is nothing wrong with a quality double-action revolver, which is often very accurate. Double-action revolvers are chambered for excellent working calibers, including the .357 Magnum. Choose a revolver that fits your hand well and has a smooth action. In my opinion, it is difficult to find a better all-around choice than the Smith & Wesson K-frame revolver. These revolvers include the Model 19 and the Model 66, each of which is chambered in .357 Magnum. With a 4-inch barrel,

Robert Alan Campbell carefully refinished this Combat Masterpiece. If you rely primarily on a long gun, a double-action revolver is a good backup, and a heavy-frame .38 with +P loads is a good choice.

this revolver is heavy enough to absorb the recoil of a powerful cartridge but light enough for long-term carry.

There are a few light revolvers that I would hate to be without. One of them is the Taurus 605 revolver. This is a lightweight revolver chambered in .357 Magnum. As such, it will handle the full range of .38 Special ammunition as well. The 605 is scarcely larger than a five-shot .38, but the barrel is heavier and the rubber grips cushion recoil from the metal frame. Recoil is there, but with a good lightweight bullet load—Cor-Bon's 110-grain .357 Magnum is ideal—the 605 .357 Magnum offers plenty of punch for the weight. This is an excellent go-anywhere, do-anything handgun.

A good double-action revolver should be in your arsenal. If you run short of your favorite handgun caliber—the .44 or .45—you can always scrounge up some .38s. Besides the .22, the .38s are the most common revolver caliber. Having a .357 Magnum around is such a good idea that even diehard 1911 men such as myself keep a Magnum revolver on hand just in case. The revolvers will take abuse and come up shooting, and while ideally we avoid poor ammunition, they will fire with ammunition that would choke a self-loader. Even underpowered or dented ammunition will work in the revolvers. These lifesavers have been around for about 175 years, so they are proven.

SELF-LOADERS

Glock

The reliable Glock currently arms 75 percent of our police officers. As such, it is hard to argue its worth. There is, however, much to be said about the Glock's safety feature. To sum it up: the Glock does not suffer fools, but then no firearm will for long. The action is easy to use, and the pistol needs minimal maintenance. The fact is, if you are primarily a rifleman and wish to use a simple, easy-to-maintain handgun, the Glock is probably your first choice. If you do choose the Glock, respect the trigger, use a proper holster that covers the trigger, and maintain strict trigger discipline.

The Glock is one modern handgun that places the tactical over the technical. It is fast into action, and the bore axis is lower than a double-action first-shot pistol. The trigger resets quickly. There is little to fault with the Glock.

1911

I can do better work with the 1911, but a quality 1911 is more complicated to use and master, as well as being much more expensive. If the question is two Glock pistols or a single Kimber, you really must ponder the question, particularly if you are also purchasing a rifle and a shotgun.

Given a quality handgun, the type matters less than the caliber in the context of the ammunition shortage. A Smith & Wesson Military and Police self-loader does have certain advantages over the Glock, including ergonomics and a manual safety. In the long run, the 9mm caliber remains the most useful for long-term shooting with limited resources. Yet with certain loads, the pistol will be an effective stopper.

This stainless 100-year-old Colt 1911 features Dan Son commemorative grips. It has been proofed with practically every personal-defense load available. Good gear, you must feed it good ammunition.

With tours in Afghanistan and Iraq behind her, Emily Campbell is showing us how to use a Colt .45. Ammunition is presently readily available for the .45, and you may not need as much as for the 9mm for obvious reasons: the .45 uses a bigger bullet and hits harder, so training sessions will probably not involve as many rounds.

When choosing a 1911 pistol, be certain you do so based on merit rather than what you think you should be carrying. You must be able to use the pistol, and this means practicing with the .45 ACP pistol. With time, practice, and diligence, you cannot be better armed with a defensive pistol. Be certain to lay in a few replacement recoil springs and at least four magazines per pistol—and 10 is better. Magazines are a renewable resource. A handful of parts is all that is necessary to keep the 1911 running for many years. With the .45 ACP, you do not have to rely on a trick bullet to produce good wound ballistics. The standard 230-grain .45 hardball will do the job well. The .45 hits hard, penetrates well, and produces a serious wound when it exits. Consider the whole picture and you will find that the total volume of such a wound is far superior to that produced by an expanding bullet that does not penetrate adequately. Just the same, with a properly designed hollowpoint bullet (such as the Hornady XTP), the .45 is even more effective. A good bet for the worst-case scenario is the Hornady 200-grain XTP loading.

As for the exact 1911 to choose, the Springfield GI pistol is probably the least expensive, yet good-quality option as you will find in the 1911 breed. The Mil-Spec is a step up. I personally have been carrying a brace of Kimber handguns for some years and have never been disappointed by their performance. The CDP, or Custom Defense Pistol, is among my favorite 1911 handguns of all time, and my gun safe includes several makers, including Colt and Les Baer. The CDP is an ideal go-anywhere, do-anything handgun. However, if you are willing to pack the extra weight, the Kimber Eclipse is an even more accurate pistol that is not as hard on the wrist due to the heavier weight.

RIFLES

.22s

A good .22 rifle is indispensable. You must have something for small game and inexpensive practice. The .22 is the first-class small-game-getter of all time.

When it comes to small-game rifles, one stands out above

Captain Matthew Campbell tries the Henry .22 rifle. This one definitely is an accurate rifle well worth its modest price.

all others: the Ruger 10/22. The rifle is dirt tough and reliable, more accurate than it has a right to be, and inexpensive. There are a host of accessories for the rifle, and Ruger offers an AR-15 type and other good variations. I have fitted aftermarket barrels that really make the AR-15 talk. Green Mountain Rifle Barrel Company offers several configurations. I have tested these and found them good. In one case, the short but stiff 16-inch barrel was the most accurate in a personal .22.

Ruger has finally introduced a high-capacity 25-round magazine that has proven reliable. I have seldom if ever found a truly reliable and practical aftermarket magazine. The 10-round magazines are adequate for most chores, but peppering predators bent on mayhem at the homestead would demand more than a handful of .22s.

The .22 has adequate penetration for personal defense if

not the requisite mass. The Winchester Super X hollowpoint is as good as any load for all-around use, inexpensive, and readily available. If you use primarily a semiautomatic AR 15 type Winchester's M22 loading is designed expressly for good cycle reliability in these rifles. It works as advertised.

AR-15s

When it comes to rifles, the terms *.223* and *rifle* are practically inseparable. I would not recommend the cheapest AR-15 rifle. Run don't walk from the parts guns. The Bushmaster carbine has enjoyed a good reputation and provides excellent value for the money. The Daniel Defense rifle is universally praised as a good, solid rifle with excellent accuracy. You cannot go wrong with the Colt. Rock River Arms has earned several government contracts for its RRA Pro Series, which it deserves.

In my opinion, the original gas impingement system is superior to the more recent gas pistol design. Whether or not you agree, there are several advantages that cannot be argued against in the older system. First, the systems are almost always compatible with each other. Gas piston systems tend to be

I am enjoying excellent results with the Daniel Defense rifle and Black Hills ammunition.

unique, which may become a logistics problem. Remember, even if you stock spare parts, there is always the chance that your rifle will break or be lost or destroyed. The next AR-15 you find should meet military specifications as closely as possible.

You should also lay in as many magazines as you can. They come in all types of quality. I have enjoyed good service with Bravo Company's offerings. The Heckler & Koch AR-15 magazines seem to be at the top of the heap; they are not inexpensive, but they are well worth the investment. (As a side note, I have no problem with the Ruger Mini-14 except that the earlier rifles were far less accurate than the AR-15. This problem has been addressed to an extent with the newer versions. Coupled

You had best have a few spare parts on hand from Brownells to go along with the ammo supply. Don't say I didn't warn you!

The author carefully sights in his Savage 99 rifle. While an older design, this one is chambered for the popular .308 Winchester cartridge.

with the lack of suitable magazines until very recently, the Ruger is not as practical for long-term use as the AR-15 types.)

If you elect to go with a .308 rifle, there are a number of proven AR-15 types. The HK91 and near copies are excellent rifles, with the domestic PTR91 a proven rifle with many good features. Be certain to use quality magazines. There are slight differences between the CETME and the HK magazines (essentially in the slot in the magagine where the magazine locks into the magazine catch). These are not insurmountable, but you must understand what they are.

Even in this day of .30-caliber AR-15 rifles, I remain loyal to the M1A1 Springfield. I desired this rifle for so long and worked so hard to afford it, I simply cannot bear to part with it for a .30-caliber AR type. The Springfield is a classic in the true sense of the word. It is accurate (with Springfield magazines) and reliable, and hits hard. You cannot ask for anything more.

Sometimes a sense of history and emotional attachment do mean something. The M1A1 is a great rifle.

SHOTGUNS

For those wishing to obtain a good, durable shotgun, there are a number of competing designs, including foreign copies of our own classic designs. Some work OK if not subjected to hard use. Others will break parts if slammed into action with a strong pump of the operating bars. The only choice in my mind is the Remington 870. Rugged, proven in police service, and capable of great versatility, the Remington is a trustworthy companion well worth its price.

A general-purpose defense shotgun with a short barrel is one of the finest problem solvers of all time. With a longer barrel, the piece is suitable for taking any flying or running game in North America, and a slug barrel will give rifle-like performance out to 100 yards. At short-range, buckshot produces a complex wound that rends and tears flesh and produces a cessation of hostilities more often than not. A 1-ounce slug is even more effective. Even my old smooth-bore Remington 12-gauge 870 will keep all the slugs on a man-sized target at 50 yards. Buckshot is more effective at longer ranges than commonly believed, but you must take time to pattern the load and master the shotgun. Buckshot travels in strings and is often unpredictable. It may be off the target at 50 yards, but if you give it a shot or two, you may find that holding to the left or right of the target produces at least a couple of hits.

Shotgun shells are often expensive and difficult to store in great number, but as this illustration of a Federal load suggests, they are often quite effective.

Winchester shotgun shells are available in a wide variety of loads. The shotgun can be a very versatile weapon.

I would not want my only long gun to be a shotgun, but by the same token, I would feel terribly limited without a shotgun on hand. As for the gauge, there is really no question. The 12-gauge is ideal. The .410 is small and easy to store but terribly expensive compared to the 12 and also far less effective. The .410 is nothing to scoff at if the range is close. The 20 gauge is a neat little shell, but again performance and economy favor the 12 gauge.

HOW MANY GUNS DO YOU NEED?

This is a good question. The essential .22 rifle, a good handgun to keep on the hip, and the shotgun may get us through

You probably will not need to stock up on as much buckshot as other munitions. One shell goes a long way.

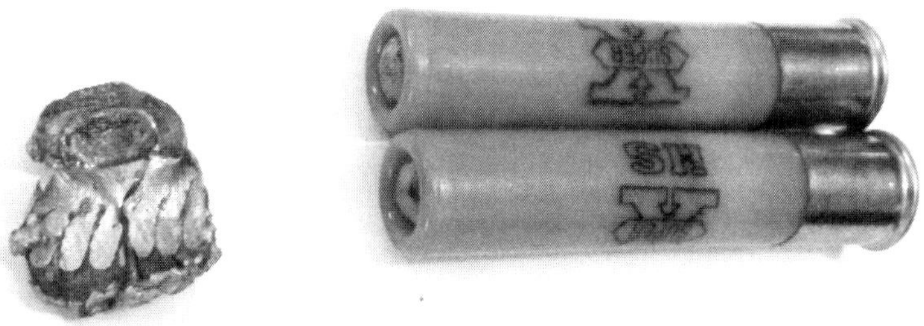

Be certain to qualify ammunition performance. The Winchester shotshell slug in .410 bore is an impressive number by any standard.

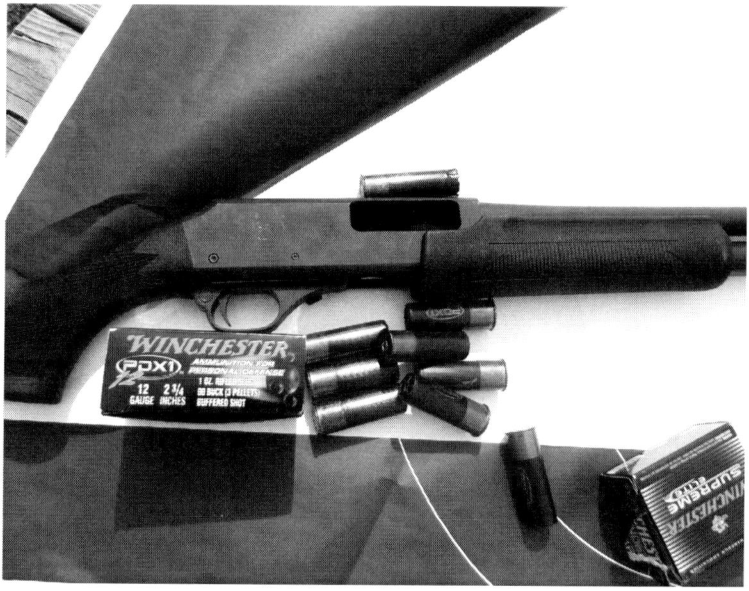

This H&R shotgun is inexpensive but effective. Winchester shot loads are among the finest available. Watch for sales and stock up.

more situations than many would admit. Ideally the .22 rifle, a good .223 or .308 depending on the circumstance, and a 12-gauge pump would be the ideal battery. Much depends on how mobile you wish to remain or how tied to the homestead you will be. The handgun remains the weapon of opportunity, at our side at all times in good and bad times.

When discussing how many guns you need, you have to address spares. One is none, and two is one. You need a spare for everything you have. Two .22 rifles, two shotguns, two rifles, and two handguns. That would be my prescription. There simply isn't any good reason not to have a spare, even it means owning a good-quality but less-impressive shotgun and rifle.

In this book, we are primarily concerned with the ammunition crisis, but we are also looking at the possibility of other prolonged crises. This means that we may need not only ammunition but support gear as well. And every shotgun and rifle needs a sling and a cleaning kit. Every handgun needs a cleaning kit and a holster—and holsters and magazines are renewable resources. You have to have spares. This isn't a program you can address overnight. It takes time and forethought. Whether you are a youngster or an old-timer, start putting together the necessary resources for the long haul. You will never regret the decision.

··· CHAPTER 17 ···

SOURCES

1. **Hornady Manufacturing Company**
   3625 West Old Potash Highway
   Grand Island, NE 68803
   Phone: 1-800-338-3220
   308-382-1390
   Fax: 308-382-5761
   www.hornady.com

2. **Nosler, Inc.**
   107 SW Columbia St.
   Bend, OR 97702
   Phone: 1-800-285-3701
   Fax: 541-388-4667
   catalog@nosler.com

3. **Sierra Bullets**
   1400 West Henry Street
   Sedalia, MO 65301
   Phone: (660) 827-6300
   Fax: (660) 827-4999
   Technical Support: 1-800-223-8799
   Toll Free Support: 1-888-223-3006
   www.sierrabullets.com

4. **Carolina Cast Bullets**
   P.O. Box 7755
   North Augusta, SC 29861-7755
   Phone: 803-426-6714
   jmo@carolinacastbullets.com
   www.carolinacastbullets.com

5. **Magnus Bullet Company Inc.**
   P.O. Box 239
   Toney, AL 35773
   Phone: 256-420-8610 Info New Number
   256-513-6633 On-line Number Skype
   256-420-8674 Fax Number
   E-mail: magnusbullets@mchsi.com
   www.magnusbullets.com

6. **Dardas Cast Bullets**
   123 East Hampton Road
   Essexville, MI 48732
   Phone: 989-450-5396
   Fax: 866-325-6525
   E-mail: Info@Dardascastbullets.com
   www.Dardascastbullets.com

7. **Space Coast Bullets**
   4250 Dow Road, Suite 307
   Melbourne, FL 32934
   Phone: 321-259-3044
   Calvin@Spacecoastbullets.com
   Richard@Spacecoastbullets.com
   www.spacecoastbullets.com

8. **Lucky 13 Cast Bullets LLC**
   1705 Abbottsford Drive
   Kernersville, NC 27284
   Phone: 336-430-5517
   don@lucky13bullets.com
   www.lucky13bullets.com

9. **Missouri Bullet Company**
   1551 SW 25th Road
   Kingsville, MO 64061
   Phone: 816-597-3204
   Website: missouribullet.com

...

ABOUT THE AUTHOR

Robert K. "Bob" Campbell holds a degree in criminal justice and served for many years as a peace officer. This is his second book for Paladin (*Holsters for Combat and Concealed Carry*) and his seventh book on firearms and shooting overall. Campbell has also authored more than 2,500 articles, columns, and reviews for major magazines, including *Gun Tests, American Gunsmith, Small Arms Review, Women and Guns, The Journal: Voice of American Law Enforcement, Concealed Carry,* and *Handloader.* His work spans subjects as diverse as history, psychology, ballistics, and police procedure. He is also the author of the horror book, *The Hunter and The Hunted.*

Campbell is the father of writers Robert Alan Campbell and Matthew Henry Campbell (author of the Paladin book *21st-Century Stopping Power*) and the nephew of songwriter and actor Gerald "Jerry" Campbell.